TOMORROW'S MATERIALS

Cover photo: Cross-section of a single optical fibre. The rings are due to changes in composition and this, in effect, confines the light beam to the inner one thousandth millimeter of the fibre.

Courtesy of Ericsson Cables,
Hudiksvall, Sweden

TOMORROW'S MATERIALS

Ken Easterling

Department of Engineering Materials
University of Luleå
S–95187 Luleå
Sweden

THE INSTITUTE OF METALS
LONDON

Book 414

published by
The Institute of Metals
1 Carlton House Terrace
London SW1Y 5DB

distributed in North America by
North American Publications Center
Old Post Road
Brookfield   VT05036   USA

First published 1988
Reprinted 1989

British Library Cataloguing in Publication Data

Easterling, Kenneth E.
   Tomorrow's materials
   1. Materials science
   I. Title
   620.1'1   TA403

   ISBN 0–901462–40–3

Library of Congress Cataloging in Publication Data

Easterling, K. E.
   Tomorrow's materials/
   Ken Easterling.
   ISBN 0–901462–40–3 (pbk.)
   1. Materials.   I. Title.
   TA403.E17 1988   87–30998
   620.1'1––dc19   CIP

Typeset by
Fakenham Photosetting Ltd
Fakenham   Norfolk

Printed and made in England by
Henry Ling Limited
The Dorset Press
Dorchester

To Edward, William, Elizabeth-Jane
and other citizens of tomorrow's world

CONTENTS

FOREWORD

As every schoolboy knows, the ages in which man has lived and progressed are named for the materials he used: stone, bronze, iron. And when he died, the materials he valued were buried with him: Tutankamun with shards of coloured glass in his stone sarcophagus, Agamemnon with his bronze sword and mask of gold, each representing the high technology of his day.

If they had lived and died today, what would they have taken with them? Artefacts of microalloyed steel, of polyether-ethyl-ketone, of yttria-toughened zirconia, of kevlar-reinforced epoxy. This is not the age of one material; it is the age of an immense range of materials. There has never been an era in which the evolution of materials was faster and the range of their properties more varied. The menu of materials available to the engineer has expanded so rapidly that designers who left college thirty years ago can be forgiven for not knowing that half of them exist. But not-to-know is, for the designer, to risk disaster. Innovative design, often, means the imaginative exploitation of the properties offered by new materials. And for the man in the street, the schoolboy even, not-to-know is to miss one of the great developments of our age: the age of advanced materials.

This concise and readable book aims to introduce the reader to the recent developments in the science of materials, and to alert him or her to the possibilities that the future holds. It is written, first, for high school leavers who are contemplating a career in science or technology; often their schooling has taught them very little about the role of materials and the opportunities that materials-related professions offer. And it is written, second, for people working in the materials or related industries who would like an overview of the potential offered by new materials. Professor Easterling has written a popular introduction to the fundamentals and likely developments of materials, avoiding the use of equations or detailed theory, and emphasising the far reaching influence of new materials on modern technology.

The text is in two main parts. The first describes the structure and properties of materials: the packing arrangements of atoms and molecules in crystals and glasses; the ways in which phases can be mixed to give composites; the cellular structures which abound in nature though they are used only on a small scale by man; and the wide range of mechanical, thermal, electrical, optical and magnetic properties that this range of structures offers. Professor Easterling's classification of material classes is novel, departing from the traditional Metals – Ceramics – Polymers; it nicely brings out the common features of all crystalline solids, and the striking differences between these and composites (enormous anisotropy, for example) or cellular solids (unusual stiffness-to-weight ratio for instance). The second part describes applications: the advances in light, strong materials for aerospace; the ways in which polymers are replacing metals in light engineering; the remarkable developments in wear resistant, high temperature ceramics; the new possibilities for information transmission offered by optical fibres, of magnet technology offered by new superconductors; and many others.

This book aims to whet the reader's appetite, to draw him or her on, to stimulate an interest in this broad and rapidly developing field. It leaves more complex and detailed explanations to other weightier, more specialised volumes. It is a book which makes the reader wish for more. And that was exactly the author's aim.

Professor Mike Ashby
Department of Engineering
University of Cambridge

PREFACE

I have written this book primarily for two groups of people. In the first group are high-school students and school leavers contemplating a career in science or technology. It is only too common to hear from them that, while they know more or less what mathematicians, physicists, or even engineers are supposed to do, they have no idea about that mysterious band who call themselves materials scientists. In the second group are those working in the materials or related industries who need a brief overview of recent advances in new materials. For both groups, I have attempted to write a popular introduction to the fundamentals and likely developments of materials, avoiding the use of equations or detailed theory. The description is far from comprehensive. Materials experts who read this book will find that a number of potentially important materials have been omitted. This is partly due to the limitations of a short and, I hope, 'readable' text. It must be admitted, however, that the materials discussed are those that particularly excite me and seem most likely to become prominent in the near future. There are no doubt others, unmentioned and perhaps undiscovered, which may well surpass them.

The text is divided into two main parts. The first, dealing with the fundamentals of materials science, gives the non-specialist an introduction to the subject and presents a novel way of classifying the various types of materials. The second part deals with applications of advanced materials and is divided into main areas of application rather than types of material. The theme developed in Part I, of avoiding the traditional classification of materials into metals, ceramics and polymers, is thus maintained in Part II, since I feel this better serves the needs of engineering and materials designers.

I should like to thank many people for helping me in the preparation of the manuscript. Of these I must mention my secretary, Sue Tuohy, for her extremely patient typing and word-processing, and Max Renner and his drawing office staff at the University of New South Wales for their excellent work on the figures. Thanks are also due to Roland Lindfors of the University of Luleå for some of the photography. I am particularly grateful to Bo Bengtsson and Hans Bentilsson, both of the

University of Luleå, to Bruce Harris, Alan Crosky, and Chris Sorrell of the University of New South Wales, and to Charles Tonkin, a pupil of Sydney Grammar School, for critically reading the manuscript and suggesting a number of useful amendments and additions.

Ken Easterling
Sydney, May 1987

PART I
FUNDAMENTALS

Four basic families of materials are considered in this first part: amorphous materials, in which atoms are arranged randomly; crystalline materials, in which atoms are stacked up in almost perfect order; composites, consisting of two or more different materials; and cellular materials, consisting of orderly stacks of hollow symmetrical cells. Each of these families includes metals, polymers and ceramics in various forms and proportions. The design of materials depends on our ability to manipulate these forms and proportions, in many cases by modifying the atomic and electron configurations through alloying and heat treatment.

Introduction

ET – extraterrestrial? No, it's short for ethylenedithio tetrathiaful-
valene, a completely new type of metal which is superconducting at
the unusually 'high' temperature of around 270°C below the freezing
point of water! Actually, much higher superconducting temperatures
have now been achieved, in a new breed of oxide ceramics. These are
just two of the new materials that we may meet in tomorrow's world
– the world of the twenty-first century. It will not be such a different
world to the one we live in today, perhaps, though it will be sufficiently changed to be noticeable.

You'll be driving a somewhat lighter car than your current one,
partly because plastics and cellular composites will have replaced
about 30% of the steel in its construction, and partly because much of
the engine will consist of lightweight nitrogen ceramics which allow it
to run at a higher temperature. These changes should please you, because the fuel consumption will be around half of what it is today.
You'll drive to work over a fibre-reinforced 'flexible' concrete bridge
of a new lightweight tubular construction. The thin concrete sections,
manufactured at the factory, will be glued together on-site using a
new super-strong polymer glue. The telephone and the electrical and
electronic functions in the car will be controlled by a tiny computer
based on gallium arsenide, or maybe even plastic (polyacetylene)
chips.

Besides new materials, you may notice quite new uses of conventional materials. For example, if the car you are travelling in happens
to be a hearse, and this is your final journey, the chances are that you
will be lying in a lightweight, foamed-polystyrene, moulded plastic
coffin. (Currently there are problems with this particular product because the material cannot yet be made to be biodegradable; however,
they do burn well and without too much smoke.)

These predictions are not the result of gazing into a crystal ball, but
are based on an appraisal of current research in leading materials
laboratories around the world. Since some ten to fifteen years usually
pass between the beginning of application-oriented research and the
production line, tomorrow's materials, based on the most promising
of today's research, may be almost commonplace by the turn of the
century.

Of course, in discussing the likely materials of tomorrow's world it is not enough to consider current research. In a dynamic world of changing political frontiers, shrinking distances, widening information networks, and growing environmental concern, attitudes to the production and use of materials will also change, probably radically. All current trends in materials research point to a much more efficient use of materials in the future. For materials which are becoming scarce this approach is obvious, and research must aim to find alternative materials to replace the old ones. But this is only a small part of the picture. Materials of the future will be produced more efficiently than ever before, with emphasis on fewer operations and much less wastage in producing the final shape. Recycling old materials is already an area of growing importance, and in this respect the non-biodegradable coffin and many other synthetic polymers will be the objects of considerable research. The Italian Government has decreed that only degradable plastic bags will be available for shoppers after the year 1991. In other words, tomorrow's materials will provide the basis of a manufacturing technology which is more energy-conscious and environmentally responsible than today's, in terms of both material production and application.

You might think that we already have quite enough materials to see us into the next century. A look at manufacturers' catalogues reveals tens of thousands of materials from which to choose. In spite of this, tomorrow's engineering designer will certainly have a great deal more from which to choose than his predecessors of today. For example, although there are currently about 15 000 different plastics available, it is predicted by scientists that this number will double by the year 2000. This is not to imply, of course, that all today's materials will continue to be manufactured tomorrow. On the contrary, changing attitudes to use and production will result in better, more sophisticated materials, designed to widen their application and provide more economical and lighter structures. Furthermore, metallurgists and ceramists are as active as their polymer colleagues in developing new products, and competition between these industries may well sharpen in the future.

The new century will witness completely new types of materials. We are now getting some spectacular insights into tomorrow's materials from the natural world, from investigations of cross-sections of dragonflies' wings, the composition of spiders' legs, the mechanics of holly leaves, the microstructure of seashells and coral, and the 'super-

4

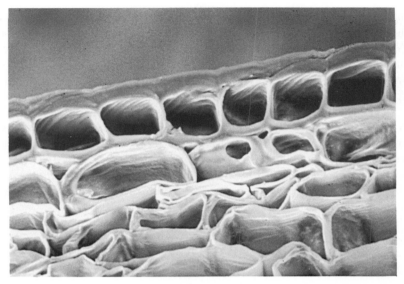

1 This electron micrograph of the cross-section of the leaf of a lily may give
   engineers clues to the design of sandwich panels   (Magnification × 700)

glue' used by mussels. The purpose of this work is to find how to apply the techniques of Nature to the design of tomorrow's lightweight cellular and composite structures. Figure 1 shows part of the cross-section of the leaf of a lily. The results of studies like this may lead to better designs for sandwich panels (these are honeycomb structures glued between hard outer sheets).

Against a background of changing attitudes, closer interaction between tomorrow's materials scientists and engineering designers is of paramount importance. This is necessary not only to enable designers to take advantage of the exciting new materials becoming available, but also to provide a basis for developing even better materials and technologies in the future. Even now, materials scientists know so much about what gives materials their various thermal, optical, electrical, and mechanical properties, that they can in effect design materials to fit a particular application in the same way that the mechanical engineer designs the shape of struts in a bridge or panels in a spacecraft.

Before getting down to the finer details of tomorrow's materials, it is useful for the non-specialist to consider briefly what it is that gives materials the properties they actually possess. For example, when is a metal not a metal? When do polymers behave more like metals? What's the difference between a plastic and a ceramic? We shall see that while the traditional division of materials into metals, polymers, and ceramics has been useful in the past, in the modern, wider approach to the structural properties and applications of materials it is better to consider materials under more general headings. When a designer comes to select a material for a particular application, all possible materials will then begin on a more or less equal footing, irrespective of their species. We shall see that a suitable division of materials can be made under the four headings *crystalline, amorphous, composite* and *cellular*. We begin with a discussion of crystalline and amorphous materials.

Crystalline and amorphous materials
Order versus chaos in the world of materials

The physicist and Nobel Laureate Sir Nevill Mott has suggested that all materials can be said to be either amorphous or crystalline. The atoms of crystalline materials are stacked up in perfect ordered symmetrical arrangements, while in amorphous materials agglomerations of atoms have little or no order or symmetry. A two-dimensional analogue would be the comparison of a regiment of soldiers on the parade ground with a mass of people in a crowded railway station. The division is not absolute, although in some ways this depends how the term 'amorphous' is defined. Many synthetic polymers, for instance, have a structure which contains a mixture of amorphous and crystal-line material, depending on how the individual polymer chains are arranged with respect to their neighbours. As shown in Figure 2, various regular or repeating arrangements can be achieved such that a sort of regimented structure is developed within a matrix of randomly arranged chains. It should not be forgotten that the chains themselves are far from random arrangements of atoms, but are highly regular and repeating arrays of molecules, as shown schematically in Figure 3. In spite of this, as a general rule polymers are regarded as amorphous materials.

With metals the distinction between crystalline and amorphous arrangements of atoms is more apparent, one being completely regular and the other quite random, as illustrated in Figure 4. Either of these

2 **The atomic chain structure of a polymer, showing random (amorphous) and ordered (crystalline) regions**

atomic structures can be obtained, the final product depending mainly upon the rate of cooling from the liquid state. Very rapid cooling (by about a million degrees per second) can effectively 'freeze-in' the random arrangement of atoms in the liquid, and such metals are called 'glassy' because of their similarity to the amorphous atomic structure of ordinary glass. However, as seen from Figure 5, the atomic structure of glass is not exactly like that of an amorphous metal; there has to be some short-range molecular order of the atoms so as to maintain the

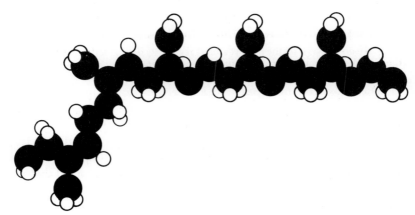

3 **A single polymer chain, showing the arrangement of carbon atoms (large black circles) and hydrogen atoms (small white circles)**

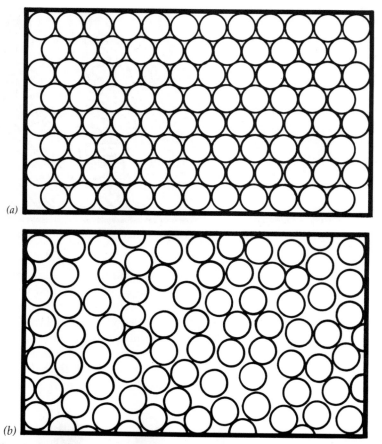

4 Comparison between (*a*) the perfect atomic regularity of a crystalline metal and (*b*) the disordered atomic structure of an amorphous metal

strict silicon/oxygen molecular structure, or 'stoichiometry', of the glass. Nevertheless, like polymers these materials are still usually classed as amorphous.

Crystalline materials are different from amorphous materials in that they can undergo plastic deformation. Plastic deformation in metals, for example, is made possible by the presence of defects called dislocations, which in effect allow atoms in one crystal plane to move relative to an adjacent plane, as illustrated in Figure 6. The movement of a large number of dislocations together brings about plastic (irreversible) deformation in a crystal; this process is illustrated schematically in

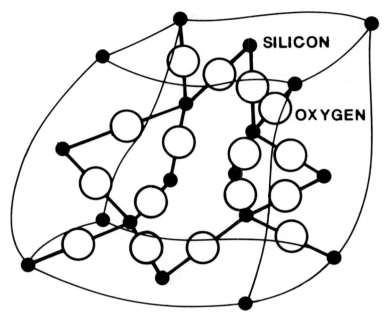

5 The random three-dimensional network of a silica glass, here likened to the distorted cubic lattice of crystalline silica

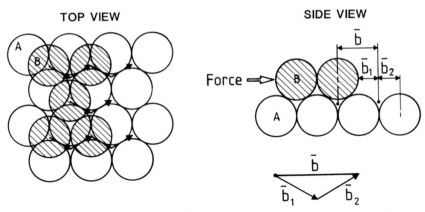

6 Dislocations allow the crystal segments of a metal to slide over one another by the movement of atoms. The vector diagram represents the individual movements of atoms in terms of 'partial dislocations'. The sum of these 'partials' is called the Burgers vector, after the person who conceived the idea

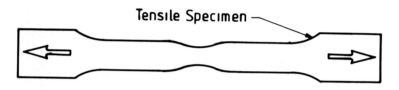

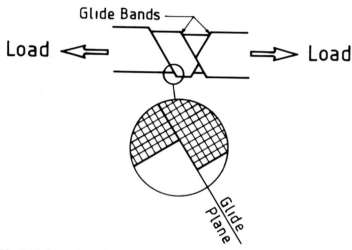

7   Plastic deformation of a metal, showing how glide planes develop along the direction of highest shear stress in a tensile specimen, a testpiece stretched along its axis

Figure 7. It can be shown that the initiation and growth of cracks is controlled by dislocation movement. In effect, a crack can be likened to a concentrated group of dislocations occurring, for example, at a crystal boundary or an inclusion (a small impurity).

While dislocations in this sense do not exist in amorphous materials, they do occur in crystalline ceramics. However, ceramics have a more complicated molecular arrangement of atoms than is found in metals, and this makes dislocation movements of the sort illustrated in Figure 7 very complicated. An example of the crystal structure of a typical ceramic material is shown in Figure 8, which illustrates the molecular relationship between oxygen and silicon atoms. When dislocations move through crystalline ceramics, the molecular arrangement has somehow to be maintained. Coupled with this, as we shall see below,

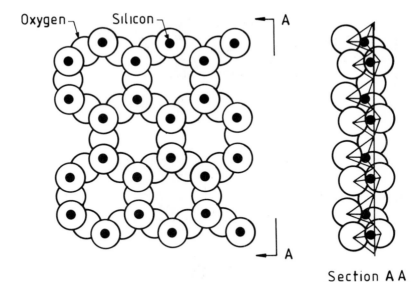

Oxygen · Silicon · A

Section A A

8 Ceramics are crystalline compounds with ordered atomic arrangements, often between metallic and non-metallic elements such as silicon or aluminium and oxygen or nitrogen. The bonding between these atoms is usually ionic or covalent, making them more directional and rigid than metallic bonds

the bonds between atoms in ceramics are more rigid than in metals, which makes plastic deformation a difficult process. Ceramics are therefore usually rather brittle. Indeed, it might be said that metals and ceramics represent two extremes of crystalline behaviour, from good ductility and high toughness (metals) to poor ductility and low toughness (ceramics). However, most ceramics are mechanically much harder than metals and usually have higher melting points, because of the powerful (oriented) molecular bonding in these materials.

An essential feature of atomic bonding in any material is the way in which electrons are shared between atoms. Indeed, a generally accepted advantage of the traditional classification of materials into metals, polymers, and ceramics is that the properties of each class of material can be related to the atomic bonding and electron configurations (*see* Figure 9). In other words, the atomic and electron configurations largely determine the characteristics of each type of material. Thus, for example, metals are typically good conductors of electricity and heat because of their regular atomic arrangement and free valence electrons. Metals are also plastically deformable, a state in which

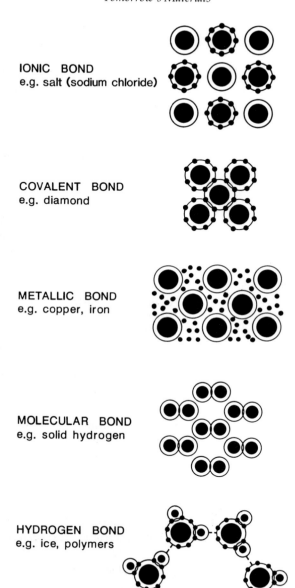

IONIC   BOND
e.g. salt (sodium chloride)

COVALENT   BOND
e.g. diamond

METALLIC   BOND
e.g. copper, iron

MOLECULAR   BOND
e.g. solid hydrogen

HYDROGEN   BOND
e.g. ice, polymers

9   The five main atomic bonding configurations in materials. The large black circles
represent atoms, and the small ones electrons. The rings surrounding atoms
represent electron orbitals. Note that the very strong oriented covalent bond
consists of shared electrons. In metals, atoms are surrounded by an 'atmosphere'
of free (valence) electrons

metals can be squeezed like Plasticene to change their shape, as in rolling or extrusion operations. This is possible because of the non-directional nature of the atomic bonding in metals, which in effect allows atoms and planes to slide over one another when subjected to sufficiently high mechanical loading (*see* Figure 7). Ordinary plate glass, a ceramic, is transparent, whereas metals are not. This is because glass is amorphous – the atoms are disordered, and there are no reflecting 'surfaces' such as the grain boundaries which make metals opaque. In addition, the bound electron shell configurations in this material allow light to pass through without significant photon scattering occurring. Glass is brittle because of its rigid atomic bonds and non-crystalline structure.

Many other ceramics are crystalline, yet are poor conductors as they lack free electrons. However, because of their crystallinity and fine-grained microstructure these materials are usually opaque. Also, most ceramics, like glass, have limited plasticity (except at very high temperatures) because of their highly rigid and directional atomic bonding. Most polymers are poor conductors (they lack free electrons) and cannot be plastically deformed in the same way as metals, again because they consist of rigid chains of atoms. However, some polymers (like rubber) can be deformed by straightening out their component chains. Other polymers in this category are polyesters, polyamides, and polycarbonates. Brittle polymers include Perspex (or Plexiglas) and polystyrene. Many polymers are transparent by virtue of their amorphous atomic structure.

In these examples, 'most' members of a class of materials have been described as 'usually' having a particular property. It is worth pointing out that current research is already producing many interesting exceptions. 'Glassy' metals have already been mentioned. Their random but relatively close-packed atomic structure provides them with unique magnetic properties. They may also be more easily superconducting – a state in which resistance to electron flow is reduced virtually to zero. However, they show little or no plasticity, and in this respect do not behave as typical metals.

An interesting recent development concerns electrical conductors made of polymers. These are made by 'doping' certain polymers with salts and iodides, which makes the electron configuration of the polymer similar to that of semiconductors. Such polymers, dubbed 'polymer metals', exhibit 100 billion times the normal conductivity of typical polymers! It has even been demonstrated that ceramics, traditionally considered to be resistors, can also be made semiconducting and even superconducting by suitable alloying. In other words, polymers and ceramics can be made to behave very much like metals in some respects.

Solidification from the melt

When a liquid is cooled below its freezing point, whether it forms into a random amorphous solid or an ordered crystalline one depends on the rate at which the liquid is cooled. Solidification usually results in an abrupt decrease in volume, because the mobility of atoms in the solid state is lower than in the liquid. (A well-known exception is the solidification of water to ice, which results in an expansion.) The volume change during the transition from a liquid to a solid in amorphous materials occurs gradually, since there is no change in the arrangement of these atoms as in crystalline solidification.

In crystalline solidification, provided the liquid is cooled sufficiently slowly a temperature is reached at which small, close-packed regular groups of atoms, or crystal 'nuclei', form. On lowering the temperature still further, these nuclei grow and finally the crystals merge together as illustrated in Figure 10. Since the original groups of atoms or crystal nuclei are not necessarily aligned with one another at the beginning of solidification (*see* Figure 10 (*a*)), the resulting solid crystal is not a single crystal but an agglomeration of several crystals – a 'polycrystal'. Each crystal is separated from its neighbours by crystal or grain boundaries (*see* Figure 10 (*c*) and (*d*)) which act as planar defects, breaking up the continuity of perfect atomic stacking of the solid as a whole. Grain boundaries are actually very important to the properties of crystalline materials, for this reason. In polycrystals, plastic deformation of the type illustrated in Figure 7 has to occur on different planes (and directions) in each crystal. Therefore, in the plastic deformation of polycrystals the whole solid has to deform in such a way that cracking does not take place between individual grains. Provided these crystal boundaries are reasonably free of impurities, polycrystals are stronger, or more resistant to plastic deformation, than single crystals. This is an important consideration in the manufacture of practical high-strength metals and alloys; in fact, it can be shown that the strength of a metal increases as the grain size decreases.

The nucleation and growth of crystals is a process that requires time. If the liquid is cooled at such a rate that there is no time for ordered

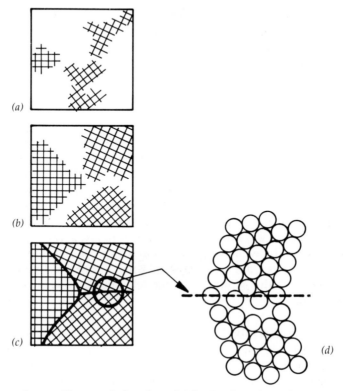

10 **How polycrystalline metals form by solidification from the melt, starting with (a)**
branched structures called dendrites, which (b) continue to grow until (c) they
meet other growing crystals to form grain boundaries (d)

crystalline groups of atoms to nucleate and grow, the liquid freezes to a
disordered amorphous solid instead. For example, when silica plate
glass solidifies the liquid first becomes viscous, rather like thick treacle.
In this state atoms no longer have the freedom to change their position
relative to their neighbours and form into ordered groups. Thus on
further reduction of temperature the silica molecules freeze into the
amorphous state of glass, because there is no systematic close-packed
stacking of atoms as in crystallisation.

   In the formation of plate glass the liquid-to-amorphous solid transition is achieved at quite moderate rates of cooling. In metals a very
rapid decrease in temperature from the melt is required if crystallisation is to be avoided. By controlling the rate of cooling from the liquid

most materials can be solidified to the glassy state, but not all materials can be made to crystallise. Synthetic polymers and natural organic compounds like alcohol, glycerol and glucose, for example, form an amorphous structure when solidifying, even at a relatively gentle rate of cooling. Whether polymers form purely amorphous (irregular chain arrangements), or mixed amorphous/crystalline structures depends upon the shape of their molecular arrangements, and not upon the cooling rate.

Solidification takes place because of thermodynamic energy difference between the molten and solid phases. This thermodynamic factor is a rather complex one and takes into account the relative bonding energies of the atoms in the liquid and solid states. It varies enormously from one material to another, which is why some materials can easily be cooled to the glassy state and others cannot. Thus to take two extreme examples, a cooling rate of up to a million degrees per second is needed to produce glassy metals, whereas a cooling rate of about a million seconds per degree was sufficient to cool the five metre diameter mirror of the Mount Palomar telescope, which is made of a silicate glass.

Crystal transformations

An important characteristic of many crystalline materials is that they can undergo solid-state phase transformations from one crystalline phase to another. A phase transformation occurs during heating or cooling when the new phase, or new atomic arrangement, has a lower energy than the old one. Thus the way atoms are packed together in iron and steel at high temperatures is different from the way they are packed at low temperatures (*see* Figure 11). The high-temperature, close-packed austenite phase of steel can also dissolve much more carbon than the low-temperature, less close-packed ferrite phase. The Japanese discovered this characteristic of iron centuries ago, and without really understanding why used this solid-state transformation to harden the edges of their swords by quenching the metal from red heat into cold water. Steel thus treated has a higher hardness because the carbon is retained in solution in the ferritic phase at ambient temperature, provided the quenching speed is sufficiently rapid to prevent the carbon atoms from diffusing and agglomerating.

Both metals and ceramics can undergo phase transformations and indeed these transformations are very important to many of the prop-

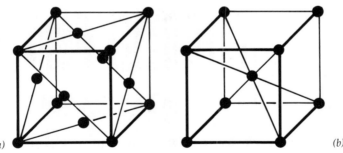

11 Two of the most common of the unit cells of metals: (*a*) face-centred cubic and (*b*) body-centred. Some metals, like iron, may be made up of both forms, in proportions that depend upon the temperature at which the iron is treated

erties of this crystalline group of materials. In contrast, polymers rarely undergo such solid-state phase transformations other than by the formation of thin, close-packed crystalline regions brought about by chain alignment. Other materials such as glass and glassy metals do not normally undergo phase transformations either, except by the process of crystallisation itself.

Composite and cellular materials

Not all materials are homogeneous solids. There are synthetic composites like reinforced concrete, and open cellular structures such as natural coral. It is worth examining these two types of material in more detail since they can differ substantially in properties from solid materials. They can also offer interesting advantages over solid materials in some applications.

Reinforced concrete and beyond

As illustrated in Figure 12, composite materials consist of at least two phases or components. This gives them properties different and in many cases superior to those of the individual phases. One of the most established of the composite materials is reinforced concrete, in which a matrix – an aggregate of cement and sand – is reinforced with steel wire or bar. The matrix is weak in tension but strong in compression, and it is also brittle, but the steel wire or bar is very strong in tension. The combination is a relatively cheap, tough material essential to the

17

RODS

PARTICLES

SHEETS

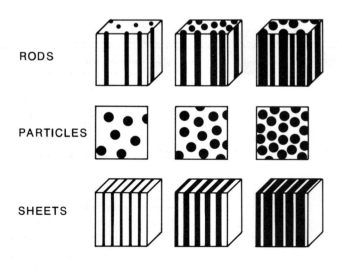

INCREASING VOLUME FRACTION ➤

12 Composites consist of at least two components, or phases, in various proportions and shapes

construction of large structures such as high-rise buildings, bridges, and oil platforms. A rather more sophisticated, yet nowadays common-place, modern composite is glass-fibre-reinforced polyester, or epoxy resin. The glass fibres are extremely strong though relatively brittle, whereas the polymer matrix is weaker but relatively flexible, so that the combination of materials and properties results in a tough, strong material. This composite has many applications including small boat structures, skis, and automobile body construction.

A great deal of research is under way to develop different types of composites and material combinations. For example, fibre-reinforced concrete with improved casting techniques which result in much stronger concrete are now being studied. Another well-known composite family is the carbon-fibre-reinforced epoxy resins used in tennis racquets and aircraft structures (airframes). While these materials are highly anisotropic (stronger in one direction than the other), they are much lighter and stronger than conventional materials such as the best aluminium alloys.

Current composite research is concerned with producing cheaper and stronger fibres that are nevertheless compatible with their sur-

Table 1 Properties of various fibres used in composites

Property	Kevlar	E-glass	Carbon	SiC	Polyethylene
Mean fibre diameter, mm	2×10^{-4}	2×10^{-4}	10^{-4}	1.4×10^{-4}	5×10^{-6}
Density, Mg m^{-3}	1.44	2.54	1.8	3.0	0.97
Young's modulus, MN m^{-2}	1.24×10^{5}	7.2×10^{4}	2.2×10^{5}	4.0×10^{4}	1.2×10^{5}
Tensile strength, MN m^{-2}	2760	3450	2070	4830	2590
Elongation to fracture, %	2.4	4.8	1.2	. . .	3.8

rounding matrix. This means that the matrix material – which can be polymer, metal, or ceramic – must be compatible and have good bonding with the fibres it supports. Traditionally, thermoset matrix composites like epoxy or polyester have been used for advanced fibre composites since they are stronger than the thermoplastic matrix composites. However, the higher ductility of the thermoplastic matrices make them more attractive for future applications. These 'advanced-composite' thermoplastics can now compete with metals in terms of toughness, corrosion and wear resistance, and even heat resistance. A comparison between some of the fibres used in advanced composites is given in Table 1. While the tensile strength of carbon fibres is not outstanding, they possess exceptionally high stiffness, or Young's modulus as defined by the steep proportionality between stress and strain. High stiffness is a particularly useful property in materials used for airframes since it is essential that large deflections of the airframe are avoided.

A drawback with composites is that because of the sophisticated production techniques required they are expensive, and so the number of applications remains fairly modest. An example of the manufacture of glass fibre reinforced tubing is shown in Figure 13. The 'laying up' of carbon fibres in different directions for a composite aircraft skin is a more complicated procedure than that illustrated in Figure 13, and is usually carried out manually. Such a manual operation, however, would be too slow and hence too expensive for automobile production. Nevertheless, the advantages of using composites for lightweight structures are such that (provided fuel costs remain high) by the turn of the century composites are likely to command a substantial share of a market traditionally held by metals, particularly in the transport sector.

13 A stage in the production of high-strength, wear-resistant glass-fibre-reinforced epoxy tubes. (Courtesy of Eivon Carlsson and ASEA PLAST, Sweden)

A leaf out of Nature's book

Cellular materials are the most sophisticated of all materials, and are in many ways among the more exciting of tomorrow's structural materials. Cellular materials form a group separate from solid materials in that their properties, such as strength, stiffness, fracture characteristics, and heat conductivity, depend essentially upon their cellular structure. Thus the way these materials react to mechanical loading, for example, is determined by their cell wall dimensions and by the shape and density of individual cells, rather than just by the properties of the cell wall material itself. Most of these materials are also anisotro-

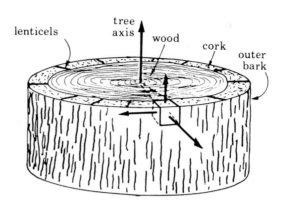

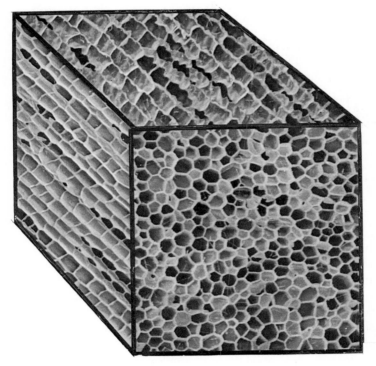

14   The microstructure of cork. There are approximately 250 000 000 of these cells in a
      single wine-bottle cork! The cellulose, suberin (a fatty substance), and waxes
      making up the cell walls of cork occupy about 10% of its volume; the rest is air.
      Its cellular microstructure gives cork many remarkable properties – for example,
      it is one of the best heat and sound insulators known. (Scanning electron
      micrograph by Ralph Harrysson, University of Luleå)

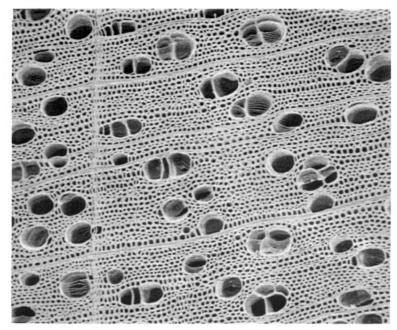

15 An electron micrograph of the microstructure of wood taken along the axis of the tree. The cells of wood are not as regular as cork, although the cell wall material (cellulose) is very similar. The elongated nature of wood cells in the direction of the tree axis makes this material highly anisotropic. Different types of tree have different cell wall thickness and different distributions of sap channels (the large cells) (Magnification × 75)

pic, which means that cellular materials react to external loads in different ways depending on loading direction. An elegant example of a cellular material is natural cork (*see* Figure 14). The tiny cells making up the structure of cork are seen to be neatly stacked like house bricks. Cork cells were first observed by Robert Hooke through the compound microscope, his own invention, in the mid-1600s. Cork was one of the earliest naturally occurring materials to be used by man, and its application as flooring, roof insulation, fishing or swimming floats, and shoes, is mentioned in Ancient Greek and Roman literature.

Another familiar cellular material is wood. As shown in Figure 15, wood consists of an array of elongated cells, the walls of which are made of a composite of several layers of spirally wound cellulose fibres in a matrix of lignin. The arrangement of these fibres is such as to give the best combination of resistance to the many types of loading to which trees may be subjected. Indeed, wood is more intricate, by far, than any synthetic cellular composite currently available.

Attempts to imitate Nature's cellular structures remain on the whole fairly crude, but in spite of this there are several examples of cellular materials in everyday use. Examples are aluminium honeycomb structures that are used extensively in modern aircraft and helicopters, and paper honeycomb structures commonly used in packaging and for internal doors in houses.

Why metal bends, rubber stretches, and glass breaks

A well-known feature of plate glass is that, while it is fairly flexible when handled with care, the impact from a cricket ball or stone thrown at it is sufficient to shatter it easily. This is rather surprising in some respects since glass is inherently a very strong material and can, in principle, withstand high mechanical loading without breaking. Indeed, as it does not contain dislocations glass is virtually as strong as the bonds between its individual atoms. Crystalline metals are also inherently strong materials in that the bonds between atoms are very strong. However, metals are much less resistant to mechanical loading than glass since they contain dislocations which can be made to move relatively easily under stress. Amorphous metals do not contain dislocations in the conventional sense, and so their strength is also, in theory, as high as that of the atomic bonds themselves. However, there is some debate about this since amorphous metals have been found to contain 'slip bands' when deformed, and this implies that some sort of dislocation mechanism may be involved. An alternative view of amorphous metals is that they are so packed with dislocations that individual dislocation line defects are not readily identifiable. Further research into the characteristics of amorphous metals is needed to clarify this, however.

The mode of fracture in amorphous polymers depends on factors other than local atomic bonding strength. For example, if you break a plastic (PMMA) ruler in two, only about 10% of the fracturing is by molecular bonds (or chains) being broken. The rest is by individual chains being pulled bodily out of the material, rather like fibre pull-

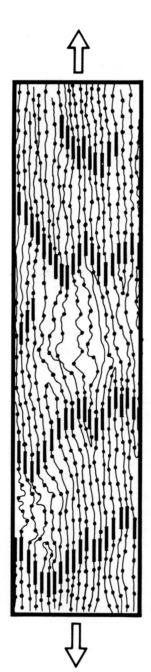

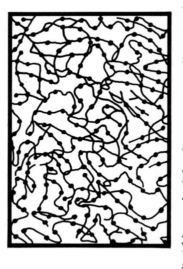

16 Stretching an elastic band can cause some alignment of polymer chains, resulting in 'crystallisation'. If this happens the elastic properties of the rubber deteriorate. The lower figure shows the polymer chains in 'relaxed' condition and the upper figure shows the stretched out chains

out known to occur in certain types of composites. In densely packed chain polymers of high molecular weight, chain pull-out occurs to a much lesser degree because the molecular chains are more tightly entangled. This aspect of polymer fracture partly explains the different speeds at which fracture occurs in these materials, in that chains require time to disentangle themselves if they are not to be broken up.

Some polymers, like rubber, are highly elastic, because of the way their polymer chains can adjust their position and shape with respect to one another. Likewise, crystalline phases in polymers are fairly rigid because there is little flexibility between the adjacent polymer chains, and the bonding of atoms across and along the chains is rigid and directional. Earlier types of elastic bands had the problem that, when stretched, polymer chains became aligned to some extent and a limited degree of crystallisation occurred, causing the rubber to become brittle (*see* Figure 16). This problem is overcome today by suitably modifying the polymer chain structure to prevent 'crystallisation' in the stretched elastic band.

The brittleness of glass is mainly due to the existence of surface defects or scratches. Glass breaks easily if a scratch is first made on its surface. The scratch acts as a stress raiser, and mechanical loading of the glass then causes the scratch to behave like a notch. When the glass is loaded the stress at the root of the scratch exceeds the bonding strength between atoms, and a crack is initiated. Cracks find it much harder to grow in metals because stress concentrations, for example at the root of surface scratches, are effectively neutralised by dislocation movement or plastic flow. Glass fibres, though, often exhibit an astonishingly high resistance to cracking when subjected to bending strains, and this is because they are relatively free of surface scratches. Plate glass can be cooled in such a way that compressive residual stresses remain at the surface, making it difficult for surface cracks to open up. Alternatively, large 'foreign' atoms can be forced into the surface layers of the glass by a process known as ion impregnation to achieve the required compressive stresses, and thereby hinder crack initiation.

So, the extremes of behaviour exhibited by different materials can be explained largely on the basis of their atomic structure. In a similar way, new materials can be 'designed' so as to have just the type of behaviour or property required.

Materials selection

In the world's production of structural materials, measured in both volume and weight, timber is in first place, ahead of concrete, with steel in third place. Plastics and aluminium lie well behind, in fourth and fifth places. Timber heads the list because in Third World countries it is still extensively used as a building material for structures such as houses and bridges. Furthermore, wood and concrete are much cheaper to produce than structural metals and plastics. To illustrate this point, the energies required to produce different materials are listed in Table 2 in gigajoules per tonne, and as an equivalent amount of oil. For example, it is seen that to produce a tonne of mild steel an energy approximately equal to 1·5 tonnes of oil would be needed. A tonne of titanium, on the other hand, requires as much as 14·5 tonnes of oil to produce. At the other extreme, it is seen that to produce a tonne of carbon-fibre composite would require more than

Table 2 The energy necessary to produce a tonne of various types of material

Material	Energy required	
	GJ/tonne	Oil equivalent (tonnes)
Mild steel	58	1·5
Stainless steel	115	2·8
Brass	97	2·5
Aluminium	290	7·5
Magnesium	415	10·7
Titanium	560	14·5
Polyethylene	80	2·1
Nylon	180	4·7
PVC	80	2·1
Rubber (natural)	6	0·15
Rubber (synthetic)	140	3·6
Wood	1	0·03
Concrete	8	0·2
Floor tiles	9	0·23
Brick	6	0·16
Glass	24	0·6
Carbon-fibre composite	4000	103

100 tonnes of oil. This, of course, is why titanium alloys or carbon-fibre composites are hardly likely to replace steel or concrete as structural materials in large fabrications.

It is a unique feature of metals that they can be used, melted down, and then used again – in other words *recycled*. This saves energy. About 40% of today's production of steel and 50% of aluminium is in recycled form, and this percentage will undoubtedly increase in future. However, there is the problem with recycling that what emerges the second time around is not necessarily the same product as the initial one. For example, scrap steel invariably contains many additional materials, such as copper from electrical wires in scrapped cars. But this has had some beneficial effects. For instance, heat-resistant nickel-based 'superalloys' were discovered by chance when materials containing 'unwanted' impurities like sulphur were found to resist high-temperature creep better than the uncontaminated version of the alloy.

There is no question that by the end of the century the amount of recycled material will have grown substantially. From an ecological point of view this is obviously a very good thing, but with high-grade metals the presence of unwanted impurities may cause difficulties. However, this problem is likely to be neutralised in future by improved techniques of analysis and better predictions of how materials behave in use.

Material properties

The properties of materials of most interest when considering structural applications are yield strength, stiffness (or Young's modulus), fracture toughness (or ability to resist crack growth) and density (mass per unit volume). In materials which will have to withstand high temperatures, the melting point is also of importance. The combination of high yield strength and good fracture toughness, or ductility, is what makes steel such an excellent structural material. When it is subjected to unexpectedly high loads (which happens more often than is generally appreciated), a steel structure does not break, but merely yields, redistributing its load over the structure as a whole. Plastics and ceramics cannot yield as they are quite brittle in comparison and this, together with the higher cost of these materials, means that structures made from them have to be designed much more carefully and accurately.

27

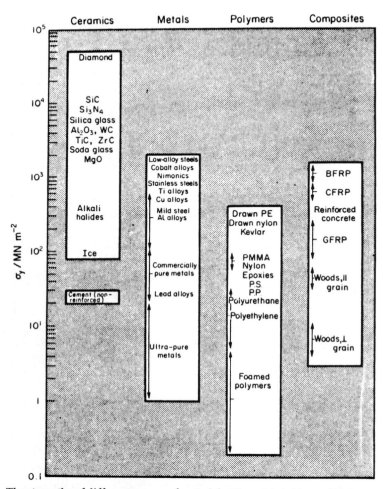

17 The strengths of different groups of materials. (From M. Ashby and D. Jones, 'Engineering Materials', Pergamon, 1980, p. 80)

Figure 17 shows the yield strength of a number of different types of materials. Not surprisingly, ceramics in general have the highest strengths, with diamond the highest of all. The strength of steels and aluminium alloys is comparable to that of fairly weak ceramics. Carbon fibre-reinforced plastics and cement are about as strong as mild steel. Of the polymers, Kevlar fibres have strengths comparable to steel, while at the other extreme the foamed polymers used in packaging and furniture have very low strengths.

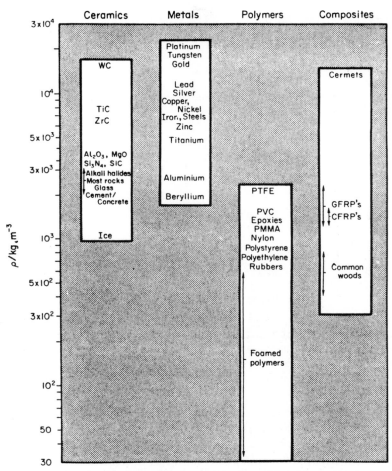

18   The densities of different groups of materials. (From M. Ashby and D. Jones, 'Engineering Materials', Pergamon, 1980, p. 53)

The densities of these classes of materials are compared in Figure 18. Ceramics are generally lighter than metals; polymers have the lowest density of all materials. Once again, composites lie somewhere in between, for the simple reason that they consist of mixtures of different materials.

When comparing materials it is important not only to consider yield strength, stiffness, and density, but to express the property in terms of the material's *specific modulus*, or *specific strength* (*see* Table 3). The

Table 3   The role of design parameter on materials selection

Material	Young's modulus, E (MN m^{-2})	Density, ρ (Mg m^{-3})	Specific Moduli*		
			$\dfrac{E}{\rho}$	$\dfrac{E^{\frac{1}{2}}}{\rho}$	$\dfrac{E^{\frac{1}{3}}}{\rho}$
Steel	210000	7·8	26923	59	7·3
Titanium	120000	4·5	26667	77	10·5
Aluminium	73000	2·8	26071	96	14·3
Magnesium	42000	1·7	24705	121	19·7
Glass	73000	2·4	30416	113	16·8
Brick	21000	3·0	7000	48	8·9
Concrete	15000	2·5	6000	49	9·5
Aluminium oxide	380000	4·0	95000	154	17·3
Silicon carbide	510000	3·2	160000	223	23·9
Silicon nitride	380000	3·2	120000	193	21·7
Boron nitride (cubic)	680000	3·0	227000	275	28·0
Boron fibres	400000	2·5	160000	253	28·2
Carbon fibres	410000	2·2	186000	291	32·3
Carbon-fibre composite	200000	2·0	100000	224	28·0
Wood	14000	0·5	28000	237	46·7

*Examples of design conditions: for E/ρ, rods in tension or short columns in compression; for $E^{\frac{1}{2}}/\rho$, beams in bending or long columns in compression; for $E^{\frac{1}{3}}/\rho$, flat plates under pressure or ceiling cladding panels.

specific modulus is expressed as a power of the modulus, the power depending on the particular design condition or criterion used, divided by density. In this approach the designer finds how a component under stress will react by calculating its strength and stiffness as a function of its weight and the type of loading applied. Thus, for rods loaded in tension it can be shown that stiffness per unit density is the correct design criterion (E/ρ). On this basis there would be little to choose between steel, titanium, aluminium, or even wood for the structural part, as is apparent from Table 3, but there could be a significant advantage in using carbon-fibre composites instead. It is surprising that for panels under uniform pressure, where the design criterion is $E^{\frac{1}{3}}/\rho$, wood would apparently be the better material to use. This is because cellular materials like wood are both strong and light, and thus have a very favourable specific stiffness in such an application.

Obviously there are a number of factors to consider when designing structures, and the specific moduli are helpful in allowing engineering designers to compare different types of materials. Also to be considered is, of course, the cost of producing a material. For example, cost would rule out the use of carbon-fibre composites in place of steel and concrete for large structures, but in sophisticated products such as tennis racquets and critical components in airframes, carbon-fibre composites may have the edge over metals.

An interesting problem is how to reduce density (and thus weight), yet still achieve a very high strength and hence obtain the highest possible specific strength in structural materials. We have seen that in

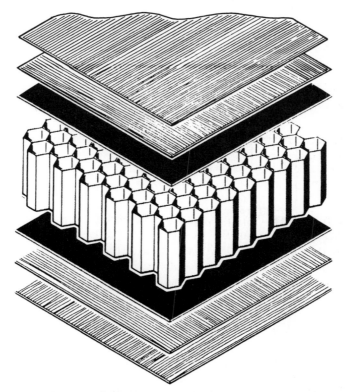

19 An example of the use of honeycomb materials in an engineering structure – part of an aircraft wing. These materials consist of a cellular structure of aluminium, polymer, or even paper, covered by one or more thin strong layers of a metal or fibre composite. The result is a lightweight material of very high stiffness, used in aircraft, skis, and door panels

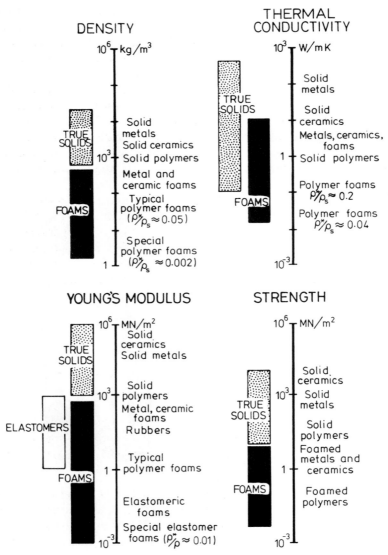

20 Cellular materials complement and considerably widen the range of properties of materials. After the work of Lorna Gibson (MIT) and Michael Ashby (Cambridge University) in a new book on cellular materials to be published by them

the natural world structural materials are usually in a cellular form; two well-known examples are wood and bone. So far we have not succeeded in producing anything like a cellular material to match wood or bone, but there have appeared a number of structural materials based on honeycombs made from aluminium. An example of the use of aluminium honeycombs for part of an aircraft wing is illustrated in Figure 19; most aircraft also have honeycomb materials for flooring and doors. Paper honeycombs are commonly used in doors in buildings, and in packaging. The very high specific moduli and strengths of cellular materials make it likely that these materials will find many more applications, particularly in the transport industry where there is a need to cut energy costs by reducing weight.

The extension to the available range of properties provided by natural and synthetic cellular materials is illustrated in Figure 20. It is seen that cellular materials complement the properties of solid and composite materials very well, in some cases (e.g. Young's modulus and density) substantially extending them. It seems likely that the future will see a much wider use of these interesting materials.

Further reading

Of the many excellent books on the fundamentals of materials science, five are particularly recommended.

R. J. Cotterill, 'The Cambridge guide to the material world' (Cambridge University Press, 1985). This book, containing many colourful illustrations, covers the entire range of organic and inorganic materials.

J. E. Gordon, 'The new science of strong materials, or why you don't fall through the floor' (Penguin Books, 1976) and 'Structures, or why things don't fall down' (Penguin Books, 1978). These two inexpensive books make for enjoyable, often humorous reading and cover the entire range of materials science.

M. F. Ashby and D. J. Jones, 'Engineering materials' (Pergamon, 1980). This book is for the more technically inclined and gives a good coverage of the interaction between materials science and engineering design.

A. J. Cottrell, 'Mechanical properties of matter' (John Wiley, 1963). This outstanding book is regrettably out of print, but it can be obtained at any good technical library. It is arguably the best book ever written on the subject of materials science, but is definitely for the scientific minded only.

M. F. Ashby, 'The mechanical properties of cellular solids', in *Metallurgical Transactions*, 1983, Volume 14A, pp. 1755–1769. This article provides an excellent overview of the relationship between microstructure and properties of both natural and synthetic cellular materials.

PART II
APPLICATIONS

This second part takes a look at some of the many exciting new materials now under development. These include new lightweight aluminium alloys and fibre–polymer composites for aircraft skins, rolled structural beams made of toughened concrete, new polymers that may soon displace metals, advanced ceramics that will revolutionise the machine-tool, electrical and automobile engine industries, the most transparent windows in the world (fibre optics), new generations of transistors, and a revolutionary superconducting ceramic which promises to change our way of life.

Structural materials

The two most important structural materials in the developed nations are steel and concrete. This is still likely to be true by the turn of the century, although plastic structures are on the way up and wood may well make a comeback. Here we look at materials used in large structures, such as ships, oil platforms, bridges, roads, railways and buildings.

Steel getting stronger

It is hard to imagine a future without steel. There is still an abundant supply of iron in the Earth's crust, and most of the alloying elements used in steel, with the exception of chromium, are also in ample supply. The current annual production of steel is around 700 to 800 million tonnes per year. While this level is likely to have decreased by some 20% by the end of the century, more sophisticated steels will continue to be developed for making structures of higher strength and lower weight. If present trends are anything to go by, these new steels will be low-alloy fine-grained materials with good weldability and toughness. The medium-strength 'clean' weldable steels for pipelines will almost certainly continue to be manufactured throughout the coming decades, on about the same scale as today, but it seems likely that aluminium alloys and carbon-fibre composites will make significant inroads on steel in areas such as automobile body production. In addition, the high-alloy steels currently used in applications where a good wear resistance is necessary are likely to be progressively replaced by ceramic-coated materials. Less stainless steel will be produced as chromium becomes more scarce, and steels of the type used for example in the chemical industry may well be replaced by titanium or coated ferritic steels. Nevertheless, the excellent all-round properties of steel will ensure that this material continues to be used for many large-scale structures such as ships, oil platforms, pipelines and buildings, and as reinforcements in concrete.

Modern high-strength low-alloy steels have a fine grain structure, which provides both high strength and good crack growth resistance, or fracture toughness. These properties are achieved by selecting a manufacturing method which gives the steel a uniform distribution

Selected wonders of the world of advanced materials 1
High strength low alloy (HSLA) steels

These materials are remarkable for the fact that they are sophisticated and yet are produced in millions of tonnes per year. They are in fact the most important material used in the construction of large structures such as bridges, ships, oil platforms and cranes. The high strength and good weldability of HSLA steels is due to their very fine grain size, typically around ten micrometres. This is achieved by very small, highly controlled alloying additions to the iron of elements that combine during hot rolling of the cast metals to produce an extremely fine distribution of tiny precipitates. These precipitates are rather complex but very stable intermetallic compounds usually based on at least two metallic components such as niobium, aluminium, titanium, or vanadium, and two non-metallic components like carbon and nitrogen. These 'low alloy additions' typically comprise only 0·15% of the total weight of the steel. The mean size of the particles is about 50 atoms in diameter and they can only be seen under a powerful electron microscope. In spite of their small size, however, these particles are very efficient in 'pinning' grain boundaries, thereby hindering grain coarsening from occurring during thermal treatments like normalising anneals or welding. The micrograph illustrates one of these tiny particles holding up the movement of a grain boundary in steel, like a fish caught in a drifting net, as observed through a transmission electron microscope.

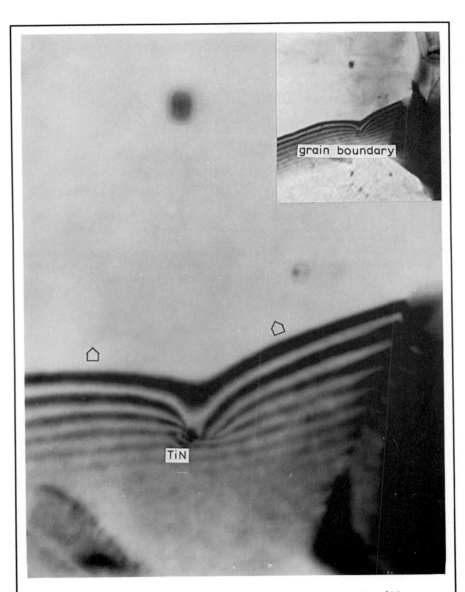

Transmission electron micrograph by Simon Ringer, University of New South Wales

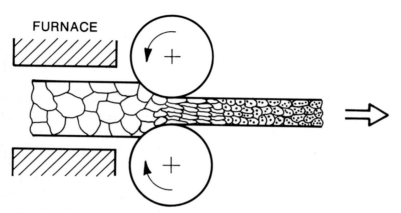

21 Controlled rolling, one of the modern techniques for producing fine-grained high-strength steels. After hot rolling the deformed structure recrystallises into fine grains which are prevented from growing by the simultaneous precipitation of extremely small carbides and nitrides. Cooling to ambient temperatures gives a very fine-grained ferritic structure with a strength of up to 600 MN m^{-2}

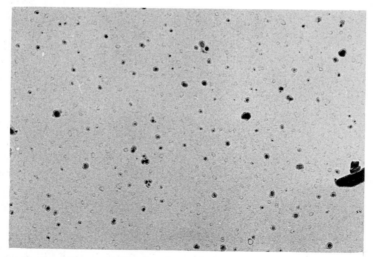

22 An electron micrograph showing fine carbides and nitrides within a single grain of a ferritic steel. (Micrograph by Jan Strid, University of Luleå)

(Magnification × 100 000)

of very small and chemically stable particles (carbides or nitrides of approximately stoichiometric composition) which effectively 'pin' or hold the grain boundaries in position. The manufacture of these very sophisticated steels has changed considerably over the past decade. In particular, the process known as controlled rolling has attracted much interest. Steel is rolled to plate at a temperature high enough for the steel to be relatively soft, but at the same time sufficiently low to cause fine carbonitride particles to precipitate out. On cooling the steel, these precipitate particles hinder excessive growth of austenite grains, so that at the end of the cooling period, following transformation to the low-temperature ferrite phase, the plate has the required fine grain size. Since yield strength is proportional to the square root of the grain size, fine-grained steels are stronger than coarse-grained steels. This hot rolling process is illustrated in Figure 21, and an example of the fine dispersion of carbonitrides present in a structural steel is shown in Figure 22.

Another way of making high-strength steel is to quench it rapidly from its high-temperature austenitic phase to ambient temperature, which produces a hard martensitic microstructure (a solid solution of carbon in iron). The steel is then tempered to produce a high-strength alloy of good toughness. This tempering heat treatment causes fine particles of carbide to precipitate in the steel, pinning dislocations and grain boundaries. For applications in which a very high ductility is needed, 'clean' steels have been developed in which there are very few interstitial alloying atoms, such as carbon and nitrogen, to strengthen the material. These steels are therefore suitable for deep drawing or severe forming operations, and in their deformed state possess very high strengths because of the high densities of dislocations produced.

Steels can be treated in this way because iron has the unique characteristic that a phase transition, from one crystal structure to another (austenite to ferrite), occurs on cooling. With a knowledge of this phase transition and the characteristics of carbide precipitation, it is possible to manipulate the microstructure of steels to produce a wide range of properties.

Welding metals

All large structures built from steel beams must be welded. This is a process in which an intensely hot electric arc is used to melt the beams locally and fuse them together usually with the aid of a 'filler' metal of

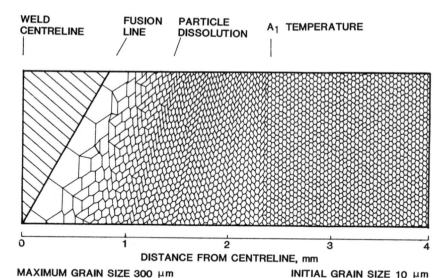

WELD CENTRELINE **FUSION LINE** **PARTICLE DISSOLUTION** **A₁ TEMPERATURE**

DISTANCE FROM CENTRELINE, mm

MAXIMUM GRAIN SIZE 300 μm   INITIAL GRAIN SIZE 10 μm

23   A computer-generated image of the heat-affected zone in a welded steel. The
intense heat from the electric arc has modified the original fine-grained structure,
producing considerable grain growth. The coarse-grained zone of the weld is
potentially troublesome in that it reduces the toughness of the joint. The fusion
line represents the boundary between solidified and base metal, and the 'A_1
temperature' label marks the place in the weld where the ferrite-to-austenite
phase transformation occurred when the plate cooled after welding. (Computed
image generated by Fred Scott, University of New South Wales)

similar composition to the beam material. In fusion welding the
temperature changes over a very wide range in a very short time,
bringing about marked changes to the microstructure and properties of
the base material. It is unlikely that fusion welding as a means of
joining metals will be replaced in the coming decades by other techniques such as gluing or diffusion bonding. Enough is understood of
the physics and physical metallurgy of fusion welding for weld cracking to be avoided. For example, welding engineers can now use a micro-
computer to predict the size and shape of the heat-affected zone for a
given welding process and material composition (*see* Figure 23). The
increase in grain size near the fusion line in Figure 23 is the main cause
of property changes since cracks grow more easily in coarse-grained
material. These microstructural changes may weaken the welded joint
further if hydrogen, picked up in the weld from moisture or dirty
surfaces, enters the steel.

Stainless steels

Corrosion is one of the most serious problems with metals such as steel. One solution has been the development of a range of so-called 'stainless' steels, each tailored to a specific application: fully austenitic steels for kitchenware and chemical plant, ferritic stainless steels for automobiles, martensitic stainless steels for hard-wearing machine parts, and so on. However, these steels have certain drawbacks in that they are rather expensive because of their high alloy content. The main alloying elements used are the fairly scarce metals chromium and nickel, which together make up some 25% of the steel. At temperatures above about 900°C the corrosion resistance of these steels fails, which can be a problem in certain types of chemical plant and in applications such as heating coils in furnaces. At present, this is overcome by surface treatments such as electrodeposition, and direct alloying or cladding with the help of heat treatments by laser beams. As mentioned above, it would be tempting to replace stainless steel with titanium alloys in chemical plant because of their superior corrosion resistance, but their cost, compared to stainless steel, is prohibitively high. Use of this valuable light metal is currently confined mainly to aerospace and sports applications, but this situation may change in the future as the metal finds more applications and becomes cheaper to produce.

Tougher concrete

Concrete is one of the most widely used structural materials, being relatively cheap and very good at bearing loads. Recent research and development has concentrated on improving the tensile strength and toughness of concrete. The results of these improvements are soon likely to be seen in concrete structures such as oil platforms, which are subject to cyclic loading.

Up to now the poor mechanical performance of unreinforced concrete has precluded its use in applications, currently dominated by metals, where good tensile strength is required, but this may well change in the near future. Recent studies have shown that the low tensile strength of cement paste results mainly from the presence of microscopic pores. A typical microstructure of reinforced concrete, illustrating the presence of pores and other defects, is shown in Figure 24. The microstructural feature of concrete that distinguishes it from

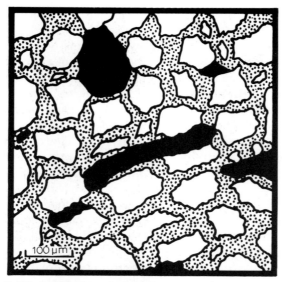

24 A schematic illustration of a typical microstructure of concrete, with considerable porosity (black areas). It is this porosity that gives concrete its poor tensile strength

ceramics is the wide range of pore sizes. It is now known that the major factor determining the strength of cement is the size of these larger pores, which result from poor particle packing and air being trapped during solidification. Indeed, if porosity is eliminated from cement then high flexural strengths, comparable to that of mild steel, can be achieved.

Various ways have been tried to reduce the porosity and improve the strength of cement. One obvious approach is to use mechanical treatments such as vibrational compaction. Cement-dispersing compounds, which effectively reduce the amount of water that needs to be added so as to give the cement paste a workable consistency, have also been used. The filling of pores with solid materials such as sulphur and resin, and the use of fine-grained cement and fibre-reinforcing materials are yet other approaches.

Perhaps the most promising development for improving the tensile strength of concrete is one in which water-soluble organic polymers are added to the cement/water mixture. The polymer/water composition alone gives a fairly stiff 'gel' which effectively forms a deformable 'dough'. Dispersed in this dough, and occupying about 60% of the

25 A schematic illustration of the microstructure of a newly developed concrete in which the spaces between the hard grains contain polymer filler. The bending strength of this type of concrete compares well with that of mild steel

volume, are cement particles, and the polymer/water gel effectively fills any excess space left between the cement particles. The resulting material is then processed by conventional plastic deformation/press-moulding techniques, such as extrusion and rolling. Once the material is formed into the desired shape, it hardens as normal inorganic hydration reactions take place within it. The final densified material thus consists of polymer-bonded, close-packed cement grains containing little or no residual porosity as a result of the filler and the type of forming processes used. The resulting microstructure is illustrated schematically in Figure 25. Other filler materials, such as silicon carbide and alumina particles or fibres, have been used in the polymer/cement mixture. In this way materials with a wide range of properties can be obtained.

It seems likely that the time is not far off when relatively high-strength beams of concrete in the form of pipes, rods, T-sections, thin sheets and even complex shapes will be in production. Complex extrusions and press-mouldings of concrete will soon be delivered from the 'concrete mill', to be assembled on site just as steel is today. Under development are new polymer-based glues capable of bonding

26 This photograph of composite wooden support beams illustrates the aesthetic beauty and structural importance of wood

together the component concrete parts, and similar to the types of polymer already used to bond the individual cement particles.

Wooden elegance

Wood is in many ways a very elegant structural material. It is cellular, with a high specific modulus which makes it more attractive for certain applications than synthetic materials (*see* Part I; material properties). Wood is still the most widely used of all structural materials in the world. This is mainly because in Third World countries it is cheap and accessible, but these attributes have led to overuse and have caused it to become quite scarce. Consequently a number of centres for wood research were established, and are now rethinking the future role of wood in structural engineering. As with other expensive and sophisticated materials, it is now apparent that more thought must be given to how wood is to be made the best use of in the future.

Materials scientists now regard wood as an advanced composite material. Its aesthetic beauty and excellent structural properties make it ideal for structural beams in buildings (*see* Figure 26). The beams in this photograph are built up from separate units in order to increase fracture strength as well as develop their overall form. Wooden units are easy to glue together – yet another advantage of this material. It is even claimed by architects that fire is no more a hazard with wood than with steel beams: as wood is cellular it has poor thermal conductivity, so even when its surface burns, its bulk retains structural strength; metal, on the other hand, being a good conductor, heats rapidly and softens.

Other wood composites under continuing development are ply-woods. There are many different varieties of plywood nowadays, providing a growing range of properties and applications. Methods of hardening and preserving wood are also receiving much attention, giving yet more breadth of application to this remarkable material.

Lightweight materials

To move anything or anybody requires energy. Whatever the mode of transport, materials used in the construction of vehicles need to combine good structural strength with high resistance to corrosion, and both they and any containers they carry must be as lightweight as possible in order to assist handling and save fuel. In principal these requirements apply equally well to cars, aircraft, railway carriages, and food and drink containers. It would seem, then, far better to use light materials such as aluminium and titanium alloys, plastics, and composites rather than the heavier steels and concrete. Unfortunately, as we have seen, most light materials like aluminium, titanium, zinc, beryllium, magnesium, and the engineering polymers and composites cost more to produce than steel, wood and concrete. This means that special requirements of cost-effective design have to be employed if these materials are to rival steel or concrete. The term 'cost-effective design' is being used more and more in conjunction with advanced high-cost materials. The implication is that components made of such materials have to be designed to use the least possible material in the most effective way. This requires far more precision in both design and manufacture than is necessary when using conventional materials.

Aluminium and the light metals

Aluminium is nowadays used for an enormous range of products (*see* Table 4), including buildings and roofing, kitchen utensils, and soft drink and beer cans. More sophisticated aluminium alloys are used for

Table 4   Applications of aluminium. (After P. F. Chapman and F. Roberts, 'Metal resources and energy', 1983, Butterworth)

Use	Percentage
Beer and soft drink cans	40
Electrical	12
Wire	5
Motor vehicles	24
Aircraft	3
Saucepans	3
Building, roofing	8
Packaging	3
Metals Industry (alloying, powders)	2

skins and frames of aircraft. Cast alloys of aluminium and silicon are widely used for automobile engine block materials, as they are strong, light, and have good corrosion and wear resistance. Titanium is more expensive to produce than aluminium, but has a higher specific modulus and is more heat resistant; in addition it is more corrosion resistant than stainless steel. Although the metal is expensive to produce, titanium and its alloys are used in a growing number of applications today including gas turbines and highly loaded components in airframes, and also in the chemical industry as a replacement for stainless steel. Zinc and magnesium are used as lightweight casting materials and for hardening high-strength aluminium alloys.

Beer cans

Beer and soft drink cans provide an enormous market for aluminium, currently swallowing at least 40% of this metal's worldwide production. Although only a relatively cheap (low-alloy) grade of aluminium is used, the production of these cans makes use of some fairly advanced materials science. Each can is stamped out from a small 5 cm diameter disc, by a deep drawing process designed to utilise the maximum possible amount of ductility that can be squeezed out of the metal. The deep drawing operation gives the elongated grain structure a certain 'texture', or common crystal orientation, rather like an aligned fibre composite. The fibrous structure, together with the very high dislocation density of the deformed material, provide the can with adequate strength against buckling when handled, in spite of its paper-thin walls. It can be shown that each cubic millimetre of can contains some hundred million dislocations, and this alone gives it considerable hardness. It is an interesting fact that about 50% of every can has experienced a previous existence as can material, as a result of aluminium recycling. Indeed, there appears to be every likelihood that this particular form of metal reincarnation may continue well into the next century!

Aircraft and spacecraft

Aluminium alloys have been used as airframe materials for around 75 years. By what is said to have been an accidental discovery, it was found that an aluminium alloy containing a little copper and magnesium, while not quite as light as pure aluminium, was considerably stronger. The alloy was called Duralumin, and several production aircraft based on this material were produced by Junkers during World War I. In spite of the fact that airframe materials have been the subject of an enormous amount of research over subsequent years, present-day aircraft are still built from a very similar type of alloy. Duralumin is actually a very sophisticated alloy in which extremely fine precipitates of particles rich in copper and magnesium, one to two atoms in thickness, harden the aluminium matrix. Recently it has been found that the properties of aluminium alloys can be improved considerably by including small additions of lithium. Lithium is the lightest of all metals, so that by using 2–3% by weight of this element the final alloy is slightly lighter than Duralumin, and it also has a higher stiffness. Although 2–3% does not sound very much, a large modern aircraft like a Boeing 767 consists of up to 80% by weight of aluminium, so tens of thousands of litres of fuel per year would be saved by replacing the present alloy by an aluminium–lithium alloy. The future is likely to see even larger lithium contents in aircraft alloys, up to 5% by weight, which will bring further improvements in strength, weight reduction, and fuel saving.

Aircraft are among the few large scale structures that are not fusion welded. Fusion welding would modify the microstructure of these sophisticated precipitation-hardened metals to such an extent that the resulting change in properties would bring about an increased risk of stress–corrosion cracking and a lower fatigue resistance. However, welding processes such as laser and electron beam welding can now produce welds with extremely small heat-affected zones, and current

Selected wonders of the world of advanced materials 2
Fibre reinforced polymer composites

These materials are remarkable because they are very strong, possess high stiffness, and yet are extremely light. As a composite, they combine the enormous uniaxial strength of, for example, fine carbon fibres and the lightweight but brittle matrix of polyester or epoxy, or other polymer matrix materials binding the fibres together. They are used as structural materials in aeroplanes and spacecraft, and in certain sporting equipment such as skis and vaulting poles. In a way, advanced composites in cellular form are synthetic copies of structures found in Nature, such as the leaves of the lily plant or the wings of the dragonfly. Lily leaves possess good mechanical strength and lightness because they are constructed in the form of tiny cells bordered top and bottom by fibrous cellulose skins. The wing sections of aircraft are similar in construction, consisting of cells of a fibre reinforced polymer or thin aluminium honeycombs, bordered by thin fibre reinforced polymer skins. Such a composite is very rigid and light in weight, and yet possesses good toughness and very high strength. Ideal characteristics in other words for use in aircraft or spacecraft. The figure is an electron micrograph of the cross-section of a fibre reinforced composite of a type used in aeroplane construction.

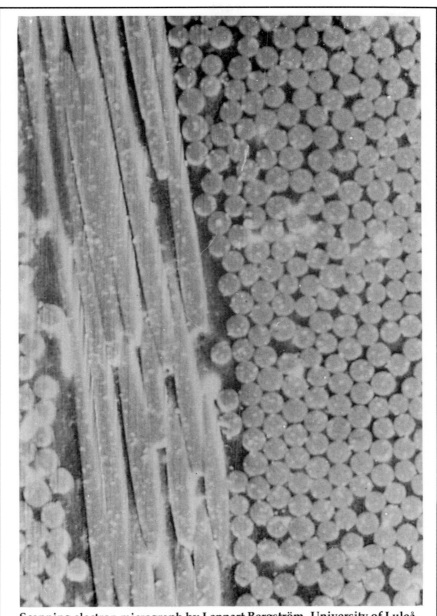

Scanning electron micrograph by Lennart Bergström, University of Luleå

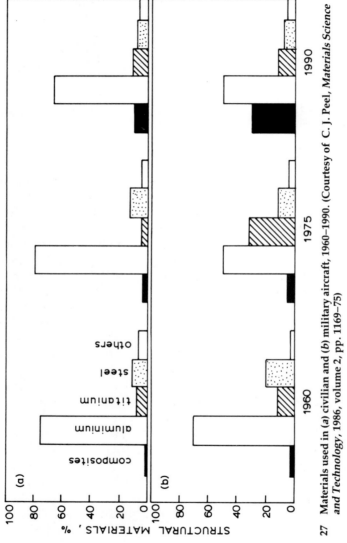

27 Materials used in (*a*) civilian and (*b*) military aircraft, 1960–1990. (Courtesy of C. J. Peel, *Materials Science and Technology*, 1986, volume 2, pp. 1169–75)

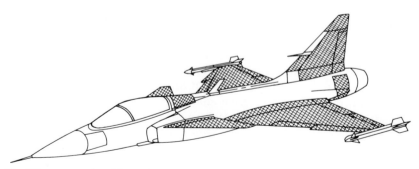

28 Fibre composites (shown hatched) are used extensively in this modern military
aircraft, a Saab JAS. Civil aircraft are rapidly following suit

research is showing that welds can be made in such a way that the
heat-affected metal precipitates and hence hardens as the weld cools.

In modern aircraft the proportion of aluminium alloys used is actually decreasing: they are losing out to composites and other materials (*see*
Figure 27). The distribution of composites in a modern jet fighter is
illustrated in Figure 28. Military aircraft normally set the trend for
developments in aviation, and civilian aircraft will probably consist of
at least 30–40% by weight of carbon-fibre composites by the turn of the
century. There are already experimental aircraft in use in which the
skin is made entirely of carbon-fibre composites, giving substantial
weight reductions over conventional metal aircraft. Honeycomb materials, consisting of aluminium cores and carbon-fibre composite
skins, are also being increasingly used in aircraft structures. The Voyager
aircraft which recently circumnavigated the Earth non-stop represents
the ultimate application of these materials in aviation – its entire structure is made from cellular polymer composites, producing a super-
light craft of high stiffness and strength.

Other composite materials being investigated for use in aerospace
structures include aluminium-containing silicon carbide ceramic fibres
and carbon-fibre-reinforced PEEK (a very tough polymer matrix material based on polyether etherketone). These toughened polymer–
matrix composites are expensive, but are less prone to cracking than
the epoxy and polyester materials currently in use. Such developments
are typical of modern materials selection, in which traditional materials
like the aluminium alloys have to compete with sophisticated new
composites and honeycombs.

Engineering polymers

Polymers form the basis of many natural lightweight materials, such as wood cellulose, starch, resins, and proteins. By studying these natural polymers materials scientists have not only been able to gain a good understanding of their molecular make-up, but in many cases have succeeded in producing synthetic versions of them. Today, polymer-based materials command a large share of the total materials market. However, many polymer scientists feel that the market for convention-al plastics is just about saturated, and that further expansion can be achieved only by developing more sophisticated products capable of pushing into areas traditionally occupied by metals. One of these areas, as we have seen, is composite materials, for which new polym-ers are being developed for use as both fibre and matrix. Indeed, polymer-based materials make up the great majority of today's compo-sites. But composites are still only a small part of a much broader class of advanced polymer products generally known as engineering polymers.

Manufacturing plastics

To produce a plastic, a whole bundle of polymers – rather like a pile of cooked spaghetti – is mixed with various fillings, plasticisers, stabilis-ers, and pigments. This mixture is shaped as desired at a temperature high enough to keep it fairly soft and then cooled into its final rigid form. Table 5 gives the more important polymer types and the plastics that are made from them. Plastics are not all polymer: polyvinyl chlor-ide, for example, used in floor tiles, contains only about 50% polymer, the rest consisting of a filler material.

Of all the plastics used today, 95% are based on as few as four main polymer types, although there are hundreds of variations of each main type. These four polymers are:

(1) polyethylene (polythene), in both solid and foamed forms, used for example in packaging, kitchen bowls and pails.
(2) polypropylene, used for example in automobile fittings
(3) polystyrene, used for example in household appliances
(4) polyvinyl chloride (PVC), commonly used for example as plastic piping in kitchens and drains.

Of the four, polyethene is the most common, with a current annual production of over 25 million tonnes.

Table 5 Polymers: structural characteristics and applications. (After H. F. Mark, special issue 'Materials', *Scientific American*, 1967, and courtesy *Scientific American*)

Polymer characteristics	Structure	Examples	Uses
Flexible and crystallisable chains		Polyethylene Polypropylene Polyvinyl chloride Nylon	Pails, pipes, thin films Steering wheels Plastic pipes and sidings Clothing
Cross-linked amorphous networks of flexible chains		Phenol formaldehyde Cured rubber Styrenated polyester	Television casings and telephone receivers Tyres, transport belts, and hoses Finish on automobiles and appliances
Rigid chains		Polyamides Ladder molecules	High-temperature insulation Heat shields
Crystalline domains in a viscous network		Terylene (Dacron) Cellulose acetate	Fibres and films Fibres and films
Chains moderately cross-linked, with some crystallinity		Neoprene Polyisoprene	Oil-resistant rubber goods Particularly resilient rubber goods
Rigid chains, partly cross-linked		Heat-resistant materials	Jet and rocket engines, and plasma technology
Crystalline domains with rigid chains between them and cross-linking between chains		Materials with high strength and high temperature resistance	Buildings and vehicles

Polymers are very versatile materials and their component chains can be knitted to produce many shapes and forms, as shown in Table 5. They can be packed together at random or interspersed with crystalline regions, depending on the method of production. These regions in turn contain linkages, called cross-links, that extend like bridges between domains. The cross-links provide extra stiffness, heat resistance and toughness. It is because of the various combinations and linkages that can be manipulated by polymer scientists that so many different types of plastics are available, and also why new forms of plastics continue to pour on to the market. Examples of new polymer types that have appeared in recent years include the tough polyether etherketone (PEEK), and the liquid crystal display (LCD) polymers. These contain very stiff chains regularly arranged, as in crystalline polymers. This 'crystalline' form is currently used as a display matrix in compact TVs and microcomputers.

Radically new polymer types based on totally new molecular forms are unlikely to appear in the coming decade or so. Apart from the scientific problems of producing them, it is now extremely expensive to launch new, unknown products into what is a fairly conservative materials market. There is probably more money to be made in developing speciality polymers which, by having better properties than some established materials, might push their way into new areas of application. One promising development along these lines has been to blend, or 'alloy', two or more different types of polymer. Today some 20% of all polymer production is accounted for by polymer blends, and injection-moulded polymer blends is one of the fastest-growing areas in the plastics field.

Polymer alloys are not to be confused with co-polymers (sequential chains of two or more types of molecule), which have been in existence for some time. They are also quite different from composite materials, in which the different materials or phases are distinguishable on a relatively coarse scale. The special characteristic of polymer alloys is that a polymer of one type is blended with that of another on such a fine scale as to be almost a single homogeneous phase – not unlike the single phase of a metal. To produce a polymer alloy an expensive polymer of one type with, say, good strength but poor formability, may be mixed with another inexpensive but brittle type which has good formability, with the aim of achieving a combination of good strength and formability. However, for this approach to work much depends upon the compatibility of the mixture as a whole. Like mix-

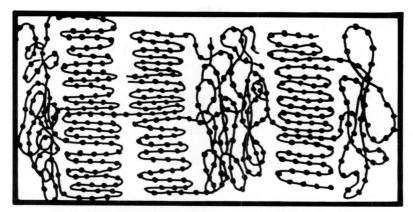

29   Higher strengths are achieved in block co-polymers in which hard crystalline particles are dispersed in a rubbery matrix

tures of water and oil, many polymers are simply not compatible with one another. As one observer has pointed out, getting two polymer molecules to alloy is like shaking two balls of wool together and hoping they will knit a scarf! The ingenuity of the polymer scientist is pushed to the limit in trying to devise blends that are both compatible and economical to produce. An alloy of this type, first produced successfully some years ago, is denethyl phenylene which goes under the tradename Noryl. It is made by alloying a polymer (PPO) which has good strength, but poor formability unless heated above 200°C, with polystyrene, which has poor toughness but good formability. The resulting blend, which can be formed at temperatures as low as 140°C, has good impact toughness and formability and is commercially successful.

There are currently dozens of new polymer alloys being investigated by polymer scientists. Among the more promising is an alloy consisting of blocks of the crystalline diphenyl siloxane and the rubbery dimethyl siloxane. This so-called block co-polymer, shown schematically in Figure 29, behaves like a thermoplastic elastomer. An attractive feature of this particular alloy is that it retains its rubbery properties over an impressively wide range of temperatures (between −50°C and +100°C). Furthermore, by varying the relative proportions of the two polymers, and with suitable heat treatment, the alloy can be made to behave as a tough elastomer or a high-stiffness material. Block co-polymers are currently being evaluated as matrix materials in carbon-fibre composites where good impact resistance is needed.

An exciting development in block co-polymers has been the production of multiphase alloys in which amorphous and crystalline phases co-exist as domains, or ultra-fine particles, in an amorphous matrix. Since the atomic bonding in the domains is ionic, the blends are referred to as 'ionomers'. These alloys are extremely tough materials in that they can withstand high-impact loading without breaking up. They are also creep resistant, and so retain their shape even when exposed for long periods to moderately high temperatures (around 100°C). Commercial examples of this type are 'Surlyn' and the ionomeric polythylene. Another example of an ionomer alloy is a polystyrene-based blend containing sodium methacrylate as the ionic constituent. With further improvements in toughness and creep properties, these light and corrosion-resistant ionomers could well begin to replace some metallic structural components within the coming decade or so.

Plastic cars and bicycles

Engineering polymers and polymer-based composites account for 20% (by volume) of the material in a modern car. While it is the more sophisticated polymers based on carbon-fibre composites that are becoming more widely used in aircraft, cheaper forms of engineering polymers are being introduced in mass-production industries like car and truck manufacture. These engineering polymers are relatively easy to produce in finished form, and are also lightweight and resistant to corrosion, but raw material costs are substantially higher than those of steel. The cost of the steel strip that leaves the rolling mill is only a tenth of its value in the finished product, but with polymers the material costs about a third of its final value in the car. So if polymers are to replace steel in the automobile industry, not only must they have satisfactory properties but also, taking all operations into account, they must be cheaper than steel (unless they have other benefits). To achieve this, a new approach to the design and production of polymer parts is necessary. For example, the possibility of producing large, aerodynamically shaped, injection-moulded polymer car bodies (e.g. from polyurethane) in just one forming operation would provide serious competition to steel car bodies, which require several stampings, spot welding, and corrosion protection treatment to achieve a comparable final shape and properties. And polymers are not being used only for car body materials: other applications include springs

30 The first all-polymer bicycle. The unconventional design exploits the good formability of polymer composites

made from an elastomer-based composite, carriage (leaf) springs of fibre glass, nylon fan wheels, wheel arches in polypropylene, and piston rings and skirts made from the new tough thermoplastic PEEK. In addition, polymer-based composites are being tested as axle materials, while moulded microcellular polymers (polyurethane foam) are already in use as bumper material.

The first all-plastic bicycle, recently produced in Sweden, is shown in Figure 30. Here again, new design and manufacturing concepts were used and the finished article, while not looking exactly like a traditional bicycle, at least gives steel bikes a run for their money! Its wheels are of a glass-fibre-reinforced polyamide and the frame is of a glass fibre/polyester, both of which are injection moulded. Even though this composite costs much more than steel, the use of direct moulding to shape the components in a single operation kept the final production cost at a competitive level.

A polymer that promises to be less of a fire risk than materials currently used in cars and aeroplanes, for example, is a new polyester/polyamide aromatic thermoplastic. Whereas most polymers are made up of heat-susceptible chains of carbon and hydrogen atoms, aromatic materials contain ring-shaped groups of atoms, making their molecules more stable at high temperatures. Indeed, these polymers are quite difficult to ignite and when they do burn they give off very little smoke. They are likely to replace existing polymers such as polystyrene or ABS as materials for internal fire-proof fittings in cars and aircraft.

Wear- and heat-resistant materials

These are areas in which considerable advances are being made in the application of new materials. Surface treatments based on alloying or direct ceramic coating are now being used for components that need to have a high wear resistance. Amorphous metals, in the form of thin surface-transformed regions or as thin foils are also appearing. As for high-temperature materials, the conventional nickel- and cobalt-based superalloys seem likely to give way to advanced ceramics with their superior heat-resisting and creep properties.

The properties of wear and heat resistance in materials are invariably related. For example, in machine tools wear resistance and thermal stability are highly interdependent, and gas turbine blades and fans must be able to withstand both heat and erosion. We now look at some modern approaches to these problems: surface treatment, powder metallurgy, and the use of advanced ceramics.

Surface treatments

The wear resistance of a material can be improved substantially by coating its surface. For example, it is estimated that the wear properties of cemented carbide machine tools and drill tips are improved by 15% if they have thin coatings of titanium nitride. These coatings are applied by a process called physical vapour deposition, carried out at temperatures of up to 450°C under well-controlled conditions. Layers thus deposited bond tightly with the tungsten carbide tool, providing superb wear resistance in the most demanding of machining and drilling operations.

Ceramic coatings such as zirconium oxide on cast iron are also being investigated for improving the wear and heat resistance of diesel engine parts. Already, ceramic coatings are commonly applied to components that will have to withstand wear, such as equipment for handling ore, by a plasma-spraying process. These techniques are currently being improved by introducing subsequent heat treatments, by laser for example, which make the coating adhere more strongly, and also densify the porous plasma-spray coatings.

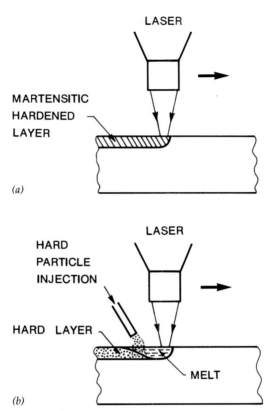

31 Surface treatment by a high-energy laser can give ordinary inexpensive metals superior corrosion or wear resistance. Two methods are (*a*) martensitic hardening and (*b*) particle impregnation

Lasers are also being used for direct hardening of high-carbon steels and other alloys. The intense heat from the laser changes the surface microstructure of the component, for example by martensitic hardening (see Figure 31(*a*)), or by alloying or particle impregnation in the laser-melted surface (*see* Figure 31(*b*)). The advantages of laser treatments lie in their accuracy and speed, and the fact that the process can easily be automated.

Surface treatments have the advantage that quite ordinary metals are in effect converted into quite sophisticated materials with greatly enhanced properties. Surface treated metals retain all their good properties, such as high toughness and ductility, and acquire in addition good resistance to wear and corrosion. They can even be made to look more aesthetically appealing! Surface treatment is just one of the ways in which materials can be tailored to suit practically any given design application or chemical environment.

Rapidly solidified metals

These materials are produced by various processes in which the alloy is melted down and then passed through a funnel or orifice into a reaction chamber for powder production, or directed onto a cooled wheel to produce ribbon in a process called 'melt spinning'. In the case of rapidly solidified powders the metal stream is met by a high velocity gas which breaks up the melt into tiny micron-sized droplets. These droplets spheroidise due to surface tension and then cool rapidly as they fall to the bottom of the chamber. In some cases the rate of cooling is so rapid that the solidified particles barely have time to crystallise and may end up in a quasi-crystalline form – something between amorphous and fine crystalline. Since the reaction chamber contains a protective atmosphere of inert gas, contamination or oxidation of the fine powder is avoided. This is an important consideration in that the surface-to-volume ratio of the powder is very high and contamination can radically affect the properties of the sintered and densified material. An example of a single rapidly solidified powder particle of an aluminium alloy is shown in Fig. 32.

The great advantage of this approach to producing metal ribbon or powders is that, even if the metal is highly alloyed, the material produced from the melt solidifies so rapidly that long range segregation or phase separation is suppressed. This means that alloys can be produced in particulate forms which would be impossible by conventional ingot casting techniques. These highly alloyed powders can then be sintered to a fully dense bulk form with little degradation due to diffusion between particles. Ideally the powder is not exposed to air but is transported directly from the reaction chamber to appropriate containers held under atmosphere, where the powder can be cold pressed into 'green' compacts and sintered.

Two areas of application in the field of rapidly solidified metals relevant to wear and heat resistant materials are high alloy heavy duty tool steels and high alloy aluminium parts for high temperature use. In the case of the tool steels, large amounts of alloying additions, such as manganese, tungsten, and molybdenum, important for producing a high density of hard intermetallic compound particles in the steel

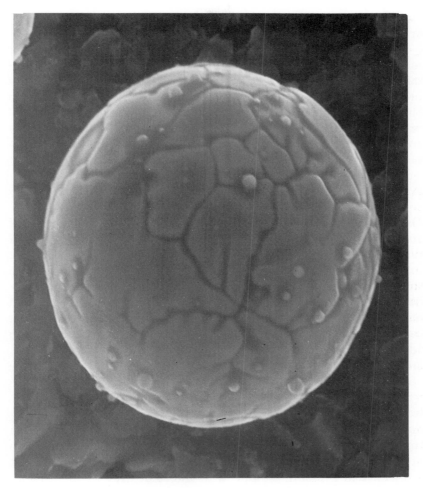

32 The microcosmology of rapid solidification showing a single Al–Mn–Cr particle.
The 'canals' in this micro-Mars are microsegregation channels. The small 'moons'
attached to the large satellite are powders picked up in the reaction chamber.
(Scanning electron micrograph by Ping Liu, Chalmers University, Gothenburg)
(Magnification × 9 000)

matrix, are quenched into the steel powders during rapid solidification. The material is then sintered and densified in a hot isostatic press,
a special autoclave in which high pressure and high temperature can
be applied simultaneously. The resultant ingots are therefore very fine
grained and homogeneous and of high toughness and heat resistance –

ideal properties for heavy duty machine tools such as used in milling operations for example.

Alloying constituents of rapidly solidified aluminium powders for relatively high temperature applications include manganese and chromium. Both these elements are capable of producing high density precipitate particles in the aluminium matrix. These alloys are found to be more stable at elevated temperatures than the conventional aluminium alloys based on magnesium, zinc, and copper additions. They are thus more suitable for parts likely to be exposed to high temperatures in certain aerospace or automobile applications.

Advanced ceramics

Oxygen, nitrogen, carbon, silicon, and aluminium are among the most abundant of the elements in the Earth's crust, atmosphere, and oceans. They are also the basic constituents of what have come to be called advanced ceramics. As discussed in Part I, these materials are sintered compounds based on carbides, nitrides, and oxides, or combinations of these compounds, and are characterised by rigid and highly directional atomic bonding which provides them with great hardness and high temperature stability. Examples of some of these advanced ceramics, together with their properties and bonding types, are given in Table 6. The impressive specific modulii of ceramics have been illustrated in Table 3.

Advanced ceramics are different from conventional ceramics, mainly in the way they are made and formed by hot pressing or sintering fine powders. In this process the powders are fused together into a solid component at a temperature high enough for rapid interdiffusion of atoms to occur at particle/particle boundaries. This process can be accelerated if pressure is applied in the same way as the powder particles of a snowball can be compacted if squeezed between the hands. The two types of ceramic are compared in Figure 33. Advanced ceramics are made by first grinding and mixing the original mineral, then purifying and filtering the powder to a fine and very uniform size, and finally densification at a high temperature. If full densification is important it is often advantageous to combine high pressure with high temperature in processes called hot pressing and hot isostatic pressing (known as HIP or 'hipping'). Alternatively, the powders may be sprayed in a plasma directly onto the surface of a metal to produce a hard coating. The resulting solid ceramic thus has an extremely fine-

Table 6   Properties of some advanced ceramics

Type	Atomic bonding	Examples	Properties
Oxides	Ionic	Al_2O_3 (sapphire) Cr_2O_3 Fe_2O_3 (hematite) MgO ZrO_2 (PSZ) $LiAl_2 SiO_6$ (glass ceramic)	Hard-wearing Good creep properties
Carbides	Less ionic Interstitial Compounds Covalent	ZrC TiC VC NbC B_4C SiC* WC	Very hard High E moduli High temperature stability Poor creep properties Used for cutting tools, abrasives, and dies
Nitrides	Covalent	BN (ambourite) Si_3N_4 AlN Sialon† TiN	Low density High temperature stability Very hard Good creep properties Used for cutting tools, gas turbine wheels, nozzles, and crucibles
Borides	Covalent	LaB_6 ZrB_2	Excellent conductor Used for electron microscope filaments Good creep properties

*SiC has properties more typical of nitrides.
†An alloy of Si, Al, O, and N.

grained polycrystalline microstructure, almost free of residual pores and defects. It is extremely hard and cannot easily be machined or cut, except by a laser; ceramics are in general much harder than metals and will easily cut steel and glass. The current revolution in advanced ceramics was made possible by the development of techniques for producing fine powders of high purity and ways of densifying them.

Like conventional pottery, advanced ceramics are extremely strong but may break easily if hit hard with a hammer: they are not as tough or as resistant to crack growth as most metals. Figure 34 shows some typical toughnesses and bending strengths for ceramics. In fact the

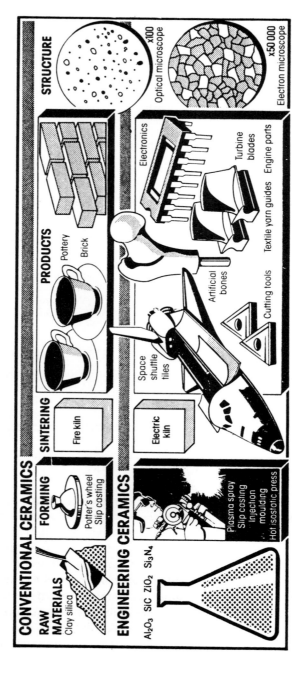

33 The essential differences between the manufacture of conventional clay ceramics and advanced technical ceramics. (Courtesy of *New Scientist*)

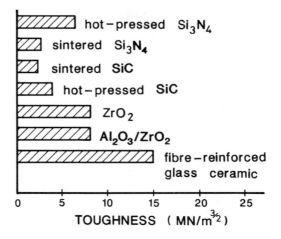

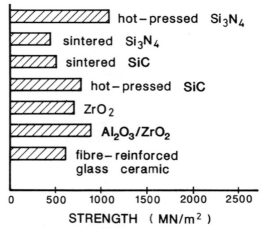

34   The toughnesses and bending strengths of some advanced ceramics. The toughnesses still lie well below those of metals

fracture toughness of ceramics, measured as the critical crack size for the onset of fracture, is only a twentieth of the values for the nickel-based superalloys. In other words, much more energy is needed to force a crack through a metal than through a good ceramic. In ceramic materials the crack usually follows the grain boundaries, where bonding between atoms is at its weakest. Materials scientists are now trying

Selected wonders of the world of advanced materials 3
Ceramic/ceramic composites

Ceramic/ceramic reinforced composites are remarkable for their great hardness and heat resistance combined with thermal shock resistance. Most important, their creep resistance or ability to withstand high stress at elevated temperature is the best of any material yet produced. Potential applications include turbine wheels, refractory components and very high wear components such as nozzles and dies. An important function of the fibres in these materials is to improve creep resistance. Advanced ceramics, being based on sintered powders, usually possess very small grain sizes. Since high temperature creep of these materials occurs mainly by atomic diffusion at grain boundaries, their creep resistance is dependent on grain size. The role of fibre reinforcement is thus to stiffen up the composite and prevent shape changes in components subject to high stress at elevated temperatures. An elegant example of a ceramic/ceramic composite is illustrated in the electron micrograph in the figure. The winding 'fibres' in this case are actually thin plates of boron nitride, an extremely light material of enormous strength and heat resistance. The fine grained matrix is SIALON (a silicon–aluminium–oxygen–nitrogen alloy), a ceramic of very high strength but relatively poor creep resistance. The composite is prepared by mixing the two materials and then hot pressing to obtain full density and good cohesion between the constituents. The whole is one of the most sophisticated creep-resistant composites yet contrived. It also possesses superb thermal shock resistance making it an excellent refractory material.

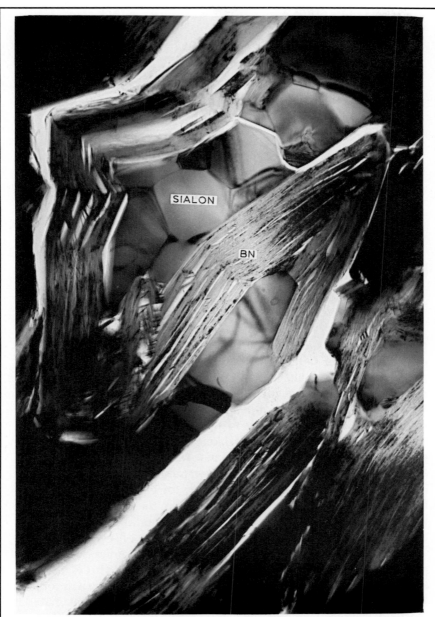

Transmission electron micrograph by William Sinclair, BHP, Melbourne

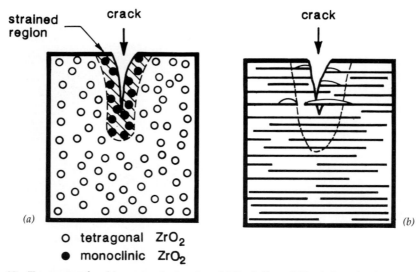

strained region

crack

crack

○ tetragonal ZrO$_2$
● monoclinic ZrO$_2$

35 Two ways of making ceramics tougher. (*a*) Partially stabilised zirconia: the dilatation of small particles by a strain-induced crystal transformation helps to close an advancing crack. (*b*) Fibre reinforcement helps to divert the crack path by debonding between fibre and matrix

to improve both the sintering and the microstructure of advanced ceramics in order to increase their fracture toughness.

One of the important factors known to govern the fracture toughness of ceramics is the residual porosity – the number of pores left after sintering. Since these pores usually occur at grain boundaries, they reduce the material's resistance to grain boundary fracture. This is one of the reasons why techniques like hot pressing and hot isostatic pressing are so effective: they minimise residual porosity or remove it altogether. Another method being examined is to introduce fine fibre reinforcements into the ceramic matrix, for example silicon carbide fibres into an alumina matrix. Toughening in these systems can occur when the main crack path is diverted by debonding of the fibre from the matrix ahead of the crack (*see* Figure 35(*b*)).

Yet another way to improve the toughness in ceramics is to exploit a phenomenon known as transformation toughening. This is rather an advanced piece of materials science, and requires the use of a ceramic which is unstable in the presence of high elastic strain. This material (partially stabilised zirconia, PSZ) transforms to its normal stable crystal structure only when severely strained, and the transformation in

36 Two twenty-first century household appliances: a knife and a pair of scissors. They have wear-resistant blades of toughened zirconia and handles of a carbon-fibre-reinforced polymer

question is 'martensitic', which means that it occurs virtually instantaneously. The metastable zirconia can be utilised in two ways. Very fine zirconia particles can be dispersed throughout a ceramic matrix of a different type, e.g. alumina, such that when a crack tries to grow through the alumina matrix and meets a zirconia particle, the particle transforms to its equilibrium phase and the resulting dilatation, or expansion, of the particle effectively closes the on-coming crack. The way this toughening mechanism helps to prevent crack growth in PSZ ceramics is illustrated in Figure 35(*a*). In an even more sophisticated approach, careful heat treatment of the zirconia causes metastable zirconia particles to precipitate within the matrix of stable zirconia. This has the advantage that a high degree of control can be exerted over the size and dispersion of the metastable precipitates. Toughened zirconia ceramics have already found use in the home as non-magnetic scissors and kitchen knives, examples of which are shown in Figure 36.

Some uses of advanced ceramics are listed in Table 7. Most advanced

Table 7   The main functions, properties, and applications of advanced ceramics. (After B. G. Newland, *CME*, January 1986)

Function	Property	Application
Electromagnetic	Dielectric	Integrated circuit substrate and packaging, electrical insulation
	Ferroelectric	Capacitors
	Piezoelectric	Oscillators, transducers, and spark generators
	Pyroelectric	Heat sensors
	Semiconductor	Thermistors, varistors, and heater elements
	Electrical conductor	Electrodes
	Magnetic	Ferrite magnets and recording heads
	Ionic conductor	Oxygen sensors, pH meters, and battery solid electrolyte
Optical	Optical condenser	Laser diodes
	Translucent	Envelopes for visible and infrared lamps
	Optical conductor	Optical fibres
	Optoelectric	Light valves and light memory element
Mechanical	Wear resistance	Shafts, bearings, seals, thread guides, process plant lining, and cutting tools
	High strength	Pressure sensors
	Thermo-structural	Engine components, welding nozzles, and jigs
	Low friction	Dry bearings and precision instruments
Chemical and biological	Corrosion resistance	Catalyst carriers, flow meters, and pump and valve components
	Chemical adsorption	Gas and humidity sensors
	Biological compatibility	Artificial tooth roots, bones, and joints
Thermal and nuclear	Refractoriness	Industrial furnace lining
	Infrared radiator	Thermal insulation and heaters
	Thermal conductivity	Heat exchangers and integrated circuit heat sinks
	Radiation resistance	Nuclear moderators

ceramics are produced for electrical and optical applications, and this is expected to continue for the rest of the century, at least. However, it is in applications requiring exceptional resistance to heat and creep that advanced ceramics appear to offer the most exciting opportunities for development.

Ceramic engines

The excellent creep resistance of ceramics makes them potentially very useful materials for diesel and automobile components, turbine blades and rotors. It has been estimated that by the year 2000, ceramics in automobile engines could have cut the world's annual energy bill by several tens of billions of dollars! This projection is based on the fact that, the hotter an engine operates, the more efficiently it runs. The use of ceramic insulation coatings, cermets, and even wholly ceramic parts in diesel engines could raise the operating temperature from about 700°C to 1100°C. This alone has the effect of improving the efficiency of the engine by almost 50%.

An advantage of using PSZ oxide coatings on engine parts is that its coefficient of thermal expansion is very similar to that of cast iron. Thin PSZ oxide coatings have already been used for a number of diesel engine parts including combustion chamber walls, cylinder liners and heads, piston crowns, and intake/exhaust parts. In gas turbines, blades made entirely of silicon nitride ceramic need no internal air cooling, so they can run at higher temperatures (and hence more efficiently) than the nickel-based superalloys used at present; ceramic bearings can operate at high speed without lubricants. Silicon nitride turbine wheels and casings for turbochargers are currently undergoing trials. Figure 37(*b*) shows an example of a hot pressed Si_3N_4 rotor mounted in a turbocharger. Ceramic turbochargers are about 40% lighter than the conventional nickel-based alloy types currently used. This means that it takes less exhaust gas to rotate them, so the turbocharger kicks into action more quickly after the engine starts, improving the vehicle's acceleration. Ceramic matrix composites are expected to replace metallic materials in jet engines by the turn of the century. The resulting operating temperature (and hence efficiency) is expected to increase from a current maximum of 1200 °C to about 1500 °C.

Ceramic tools

Advanced ceramics are ideal wear-resistant materials. Even traditional wear-resistant materials like cemented carbides (tungsten carbide/cobalt compounds) are likely to be augmented by ceramics or metal-coated ceramics in the not too distant future, particularly now that tungsten is becoming scarce. Ceramics are already widely used as refractory furnace linings, but many new applications of wear-resistant ceramics are currently being investigated: coatings for machine tools, fully ceramic machine tools, coatings for draw rolls in textile machinery, rolls and dies in metal-forming plants, and so on.

(a)

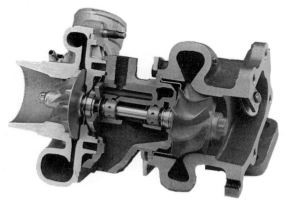

(b)

37 Examples of advanced ceramic production: (*a*) a turbo rotor together with various machine tools and ball bearings, all of hot pressed silicon nitride; (*b*) the ceramic turbo rotor in position in the turbine. (Courtesy of NGK, Japan)

The use of ceramics in machine tools currently represents only about 2–3% of the total machine tool market, but by the end of the century it is thought that this will have increased to as much as 30%. The ceramics of interest here are Al_2O_3, Al_2O_3/TiC, sialon, and Al_2O_3/SiC fibres. Typical cutting speeds for hot isostatically pressed sialon, for example, are around 2000 rev min^{-1}, compared with 800 rev min^{-1} for TiN-coated WC and 1000 rev min^{-1} for Al_2O_3/TiC tools. Ceramic Si_3N_4 bearings and machine tools are illustrated in Figure 37(*a*).

Bioceramics

A new and growing area for advanced ceramics is biomaterials. Traditionally, metals and polymers have been used for prosthetic or replacement devices in medical and dental surgery. In recent years, however, dozens of different ceramics have been investigated as potential implant materials. What makes them particularly suitable are their superior wear and erosion characteristics compared with other materials. For example, orthopaedic surgeons have found that the replacement of damaged bone by an advanced ceramic implant is advantageous in several respects. Not only is the high strength-to-weight ratio (the specific modulus) of considerable advantage, but advanced ceramics, during densification by a powder sintering process, can be deliberately made porous, and this enables regenerating bone to grow into and bond with the implant. Furthermore, ceramics do not corrode as do other materials used for this purpose, and they are not too rigid, having enough flexibility for body movement. The ceramics of most interest as biomaterial implants are Al_2O_3, Si_3N_4, and a complex 'bioglass' based on SiO_2. Current and potential markets for bioceramics are given in Table 8. It is thought that by the end of the century the value of this market could almost treble in value to some 10 billion dollars.

Spaceage ceramics

A novel application of heat-resistant ceramics is the well-known tiling used to protect the space shuttle during re-entry into the Earth's atmosphere. As illustrated by the electron micrograph of one of these tiles (*see* Figure 38), the tile consists of an open cellular microstructure of extremely fine coated silica fibres. The fibres are so loosely packed that the tiles consist of 95% air, making them as lightweight as cotton wool. Cellular materials are very poor conductors of heat. Trying to

Table 8 Current and predicted applications and costs of bioceramic materials. (L. L. Hench, *Advanced Ceramics Materials*, January 1986)

Application	1986			1996		
	Number produced (000)	Unit cost ($)	Market ($ million)	Number produced (000)	Unit cost ($)	Market ($ million)
Total hip	300	600	180	400	700	280
Total knee	150	400	60	200	500	100
Ankles/elbows/shoulder	50	500	25	700	600	420
Finger	400	50	20	600	100	60
Fixation pins/plates	1000	50	50	1200	60	72
Tooth implants	300	500	150	1500	250	375
Periodontal treatment	200	20	4	1000	50	50
Ridge augmentation	100	20	2	2000	40	80
Ridge maintenance	250	10	2·5	15000	20	300
Mammary prostheses	400	100	40	500	120	60
Intraocular lenses	1000	400	400	1500	750	1125
Middle ear prostheses	30	150	4·5	35	200	7
Cochlear prostheses	0·5	10000	5	10	10000	100
Sutures	17500	2	35	18000	2	36
Facial augmentation	6	200	1·2	10	250	5
Myringotomy tubes	300	30	9	500	50	25
Pacemakers	190	1500	285	200	2000	400
Heart valves	40	2000	80	60	2500	150
Arterial prostheses	250	200	50	500	250	125
Ventricular assist	0·1	7500	0·75	0·5	10000	5
Vascular bypass	500	?	?	1000	?	?
Total: USA			$1403·95			$3775
(World)			($3509·87)			($9437)

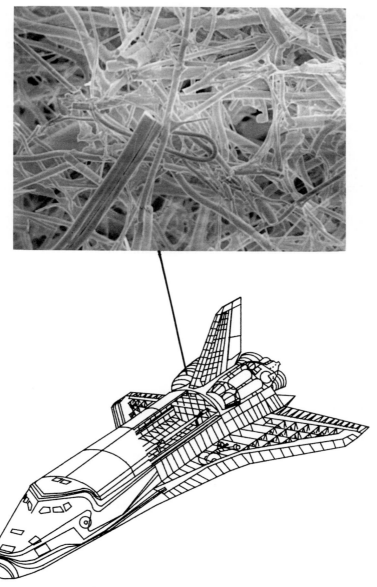

38   An electron micrograph of the open cellular microstructure of the heat-resistant
      tiles used on the space shuttle. Heat conduction is virtually confined to the open
      space between the silica fibres, and is therefore extremely low. (Micrograph by
      Anthony Bourdillon, University of New South Wales)

heat them up is rather like attempting to warm a cauliflower in the oven – the surface soon gets hot but the middle stays amazingly cool! These tiles are probably the most heat-resistant materials ever produced, being both low-density cellular and ceramic; during re-entry their outside temperature can reach 1500°C (hotter than molten steel), yet they continue to do their job.

New advanced ceramics are among the most exciting of tomorrow's materials. Although complex and expensive to produce, they are made from the most abundant of the Earth's natural materials. Materials scientists are provided with perhaps their greatest challenge in ensuring that these attractive materials, with their problems of processing and brittleness, can be fully utilised in the future. Engineering designers also need to adopt a different approach to their use since the size and effectiveness of components based on these ceramics must be far more highly optimised than those using cheaper conventional materials. In some respects quite new design philosophies are necessary. However, the potential payoff in terms of performance, weight saving, length of life, and reliability is enormous.

Optical materials

Research into these materials is currently very active, and has led to marked improvements in strength and texture, and to greatly improved coloured plate glass. Selective absorbents in glass are already widely used to filter ultraviolet light in windows in high-rise buildings and in sunglasses. In telecommunications, the use of fibre-optical materials offers huge advantages over conventional metallic materials in terms of performance (information carrying capacity) and bulk weight of the cables. These advantages are likely to increase substantially as signal losses of the optical fibres are further reduced, allowing signal transmission over hundreds of kilometres without intermediate amplification. Material for solar cells is becoming progressively cheaper and more efficient and may provide the key to cheap, 'clean' energy, even for Third World countries.

Stronger glass

Amorphous materials like glass effectively inherit the composition of the melt from which they form, so it is an easy matter to add various elements to the melt to obtain glass with whatever properties and colours are required. For window glass, additions such as sodium or potassium occupy some of the silicon sites in the glass lattice and thereby lower the binding energy between atoms in the glass molecules. This results in a material that is easier to form when hot. Additives can also be selected to help inhibit crystallisation and to improve the fluidity of the glass in its viscous state. Additives like potassium concentrate in the surface layers of glass (e.g. by ion implantation), where the presence of the large potassium atoms creates compressive stresses and so improves the glass's resistance to fracture. Alternatively, nitrogen atoms can be introduced into the glass where they partially substitute the oxygen atoms of the silica. The end-product is a glass of high through-thickness strength due to the very high covalent bonding strength of the Si–N molecules. These techniques are known as 'chemical' strengthening of the glass. Similar effects can also be achieved by thermal treatments, such as quenching the hot glass in air

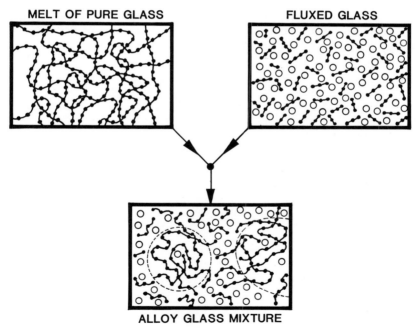

MELT OF PURE GLASS   **FLUXED GLASS**

ALLOY GLASS MIXTURE

39 Tough alloy glasses are produced from phase mixtures, in much the same way that polymer blends are produced. (From R. J. Charles, 'Materials', *Scientific American*, 1967)

Selected wonders of the world of advanced materials 4
Optical fibres

These materials are remarkable because of a sophisticated processing route which provides fibres with enormous
information-carrying capacity but very small losses in light
intensity. Thin layers of different compositions are deposited
on the inside of a silica tube using a deposition process from
flowing gas. A computer monitors the composition of the gas
so as to continuously modify the deposited layers of material.
Following this, the tube is hot drawn into long, solid fibres of
about a tenth of a millimetre diameter. In this form the graded
change in composition produced by the deposition process
effectively controls the refractive index of the glass fibre, such
that the transmitted light beam is confined to the inner
thousandth of a millimetre of the fibre. In this way, the loss of
light intensity due to 'bouncing' within the fibre are reduced
almost to zero. More recently, fluoride fibres have been produced with such low losses that light can be transmitted over
hundreds of kilometres without need of reamplification. The
figure illustrates a silica fibre, and its corresponding graded
composition is shown in the form of concentric rings within
the cross-section of the fibre.

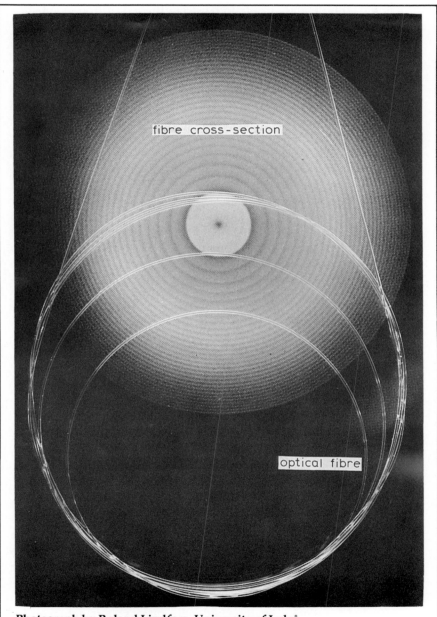

Photograph by Roland Lindfors, University of Luleå

or a liquid. If bulk glass needs to be strengthened, nitrogen substitution or fibre reinforcement can be used.

In recent years work on the effect of additives on the properties of glass have established that, besides strength, electrical and optical properties may also be improved. In its immiscible state, the glass consists of an extremely fine interdispersed phase mixture. The way in which the two phases separate out during the liquid to viscous solid transition is characteristic of the precipitation of chain structures in polymerisation. The result, illustrated schematically in Figure 39, is not unlike the structure of polymer alloys. Like polymers, glass alloys can even be produced in partially crystallised form by using appropriate additives and heat treatments. Besides having better formability, these alloy glasses are found to be stronger than ordinary glass.

Alloying or doping can be used to bring about electrical conduction in silicon glass, just as doping silicon makes it semiconducting. It seems likely that alloying glass to change its properties will continue to be an important area of development in the future, providing much stronger and more colourful glasses as well as opening up the possibility of applications in the semiconductor field.

Fibre-optics

The flow of light along fibres can be likened to the flow of water through a pipe. Advanced laser and amplifier systems have made it possible for fibres to transmit light over distances of 200 kilometres without the need for a 'boost', and such fibres are now being tested. By the end of the century the 'window thickness' of fibre optics will have reached 500 kilometres. By then the speech-carrying capacity of individual fibres will have increased enormously, and fibre optics will connect individual homes to international networks, bringing into sight high-resolution, worldwide TV and 'videophone' networks.

Fibre-optical telecommunication systems work as shown in Figure 40. When someone speaks into a telephone, the voice sounds are converted first into electrical signals and then into a sequence of light pulses by a laser beam. These pulses are transmitted along the fibre to the receiving end, where they are converted back to normal sound. For long-distance calls the light pulses have to be confined to a thin cross-section of very pure glass at the centre of the fibre so that there are no losses in light intensity by repeated reflection. This is achieved by making the fibres in such a way that the glass's composition, and hence

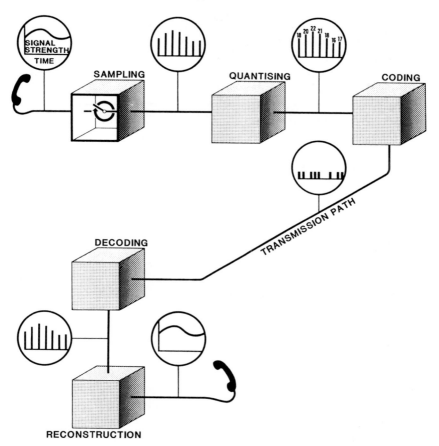

40 The application of fibre-optical systems in telecommunications. The very high
frequency of the light waves allows thousands of telephone conversations to be
carried simultaneously by a single fibre less than a tenth of a millimetre in
diameter. (Courtesy of the Science Museum, London)

refractive index, varies from relatively impure glass at the outside of
the fibre 'bundle' to a fine central fibre of high purity. The change in
composition over the cross-section is carefully controlled by a process
of vacuum deposition, so as to optimise the refractive index of the
light-transmitting fibre. An example of such a graded composition
fibre is shown in the figure on page 81. Light rays are confined to the
central thousandth part of a millimetre of the fibre; light appears in
effect to disobey the laws of optics by following the contours of the
fibre as it winds its way from one location to another. The loss of

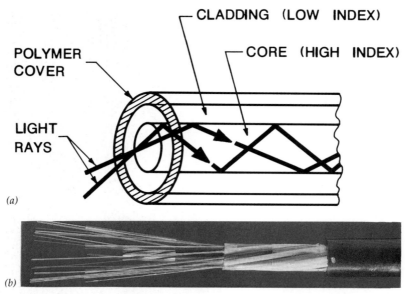

CLADDING (LOW INDEX)

POLYMER COVER

CORE (HIGH INDEX)

LIGHT RAYS

(a)

(b)

41 (*a*) How light waves bounce along a fibre enclosed in a sleeve of high refractive index. The secret of modern fibre optics is to reduce these reflections to a minimum, thereby increasing the efficiency of transmission. (*b*) A fibre-optical cable

light-transmitting efficiency suffered by modern fibres amounts to only 0·2 decibels per kilometre. This means that light can travel 20 km along the fibre before losing half of its original intensity. This efficiency will certainly be improved in the future, and 0·1 decibels per kilometre should be achievable even with present-day technology.

An important advantage of glass fibres over conventional copper conductors in telecommunications is the extremely small diameter of the fibres compared to their metallic counterpart, together with their far greater load-carrying capacity (*see* Figure 42). Typically glass fibres have a diameter of less than a tenth of a millimetre, so even when clothed in their polymer jackets, many more fibres can be fitted into a cable of a given size than can copper wires. Furthermore, unlike metallic cables, fibre-optical carriers are not affected by electromagnetic disturbances. This has, for instance, opened up the possibility of suspending the cables over railway lines. Already in the USA networks of fibre-optical cables are linking individual homes to the national TV networks, and similar fibre-optical networks for 'videophones' are projected. With plans for transoceanic fibre-optical cables progressing

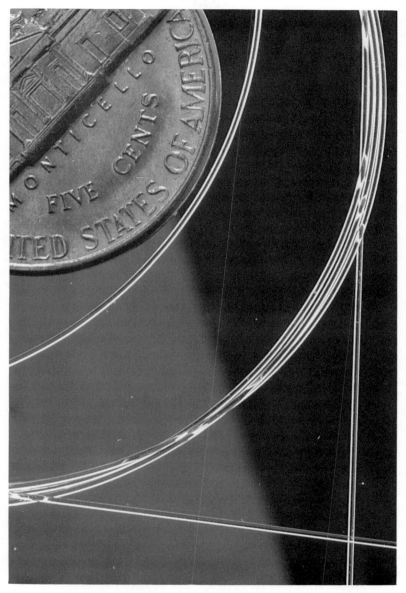

42　A single optical fibre as illustrated can carry simultaneously some 8000 telephone conversations or 60 TV video transmissions. These not-so-delicate silica threads hold the key to worldwide TV and vision telephone networks by the turn of the century. (Photograph by Roland Lindfors, University of Luleå)

well, these advanced telecommunication systems will shortly span the world.

Recent research suggests that scientists are not satisfied with the very low optical losses of glass fibres, and are already investigating other new materials with the object of improving light-transmitting efficiency still further. For example, fluoride fibres, with complex compositions at the core, have been produced and found to exhibit only a tenth of the losses of the best silica fibres.

Solar cells

Like fibre optics, solar energy conversion is likely to become a very important technology. It promises to be a cheap and 'clean' energy source, particularly suitable for Third World countries where other energy sources, based for example on oil or atomic power, are expensive.

One of the best solar cell materials is crystalline silicon, the material used for example in space probes and household solar converters. Crystalline silicon has a higher conversion efficiency than amorphous silicon, although it is an expensive material. While high costs may be acceptable for sophisticated space applications and luxury homes, the cost of large ground-based solar energy installations using crystalline silicon would be prohibitively high. A cheaper alternative material for large-scale solar cells may be amorphous silicon. It can be cheaply produced in the form of very thin coatings evaporated on to a suitable substrate material. Even very thin evaporated films, of the order of a thousandth of a millimetre thick, are able to convert solar energy fairly efficiently.

Current research is aimed at improving the efficiency of amorphous thin-film silicon cells, for example by employing layered structures rather like those found in modern fibre optics. In this way, it is envisaged that giant solar energy installations receiving sunlight reflected down from orbiting satellites will meet the energy needs of whole communities and even industrial plants. Other ambitious projects being considered include the possibility of achieving solar energy conversion in space, where the process would be much more efficient. The energy could then be beamed to a receiving station on Earth by using microwave transmission.

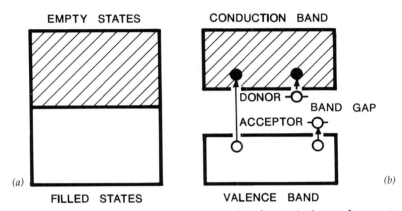

EMPTY STATES
CONDUCTION BAND
DONOR
BAND GAP
ACCEPTOR
(a)
(b)
FILLED STATES
VALENCE BAND

43 The energy bands of (*a*) a metal and (*b*) a semiconductor. An input of energy to the semiconductor in the form of light or heat causes a valance (free) electron to jump up to the conduction band. The role of graded impurities, or dopants, is to provide 'carriers' with energies within the 'forbidden' gap. Some dopants donate electrons to the conduction band, while others accept electrons, creating 'holes' in the valence band. Doped silicon chips are based on this principle

Electronic and magnetic materials

Just as steel changed the world during the industrial revolution, so the semiconductor has had a similar impact in the 1970s and 1980s. 'Silicon Valleys' are now found in many places other than California, and the importance of silicon to our chip-shaped world is likely to continue for some years to come. However, new semiconducting materials are appearing on the horizon, some of which will certainly soon begin to steal some of the limelight from the ubiquitous silicon. A milestone in this field has been the discovery of the 'high-temperature' superconductors based on copper oxide and capable of operating well above liquid air temperatures. These remarkable ceramic materials herald a whole new generation of devices – miniature high-speed computers, super-magnets, levitating trains, and compact nuclear magnetic resonance scanners for detecting tumours.

A chip-shaped world

Today's semiconductors are based mainly on crystalline materials, for example silicon or gallium arsenide, doped with boron, arsenic, or phosphorous. As illustrated in Figure 43, the role of these dopants is to pass electrons to the host material or accept electrons from it, causing it to become conducting.

As shown in Figure 43(*a*), metals are conducting without the need for alloying additions or dopants because of its partially filled conduction band, in which electrons jump easily from the filled to the empty parts of the band, where they are free to move from ion to ion. In Figure 43 the metallic conduction band is represented by both empty and filled states. Because there is no band gap, very small changes are needed for electrons to move from the filled to the empty state. With semiconductors, however, an input of energy in the form of light or heat is needed to cause electrons to jump over the band gap and bring about conduction. Silicon semiconductors have quite a wide band gap, and the energy required to bring about conduction is therefore quite high. This causes the chip to warm up, and in large computers like the Cray II efficient cooling systems are essential. Other semiconductor materials such as gallium arsenide have smaller band gaps, so less energy is needed to activate them. Materials such as these will provide more efficient semiconductors in the near future.

The technological importance of semiconductors is that the contact between a semiconductor and a metal (a full conductor) acts as a rectifier, enabling electric current to pass more easily in one direction than the other. In computing, for example, the ability to provide low-energy switches, in which small rectifying junctions are used to turn current on and off extremely rapidly, is vital to the various operations computers are required to carry out.

A problem with crystalline materials such as silicon or gallium arsenide which are to be doped is that they must be in an extremely pure form, otherwise unwanted defects such as impurity atoms or even dislocations tend to disturb or interrupt the flow of electrons. It is worth taking a closer look at the materials science behind semiconductor production, particularly silicon chips.

Silicon chips

The raw material for silicon is quartz or silica, which is melted and refined to produce metallic polycrystalline silicon. This metal is then melted in a crucible, and a 'seed' crystal is used to grow a single crystal of silicon. This must have a minimum purity of 99.999 999 9%. The next step is to cut the crystal into thin wafers, about a quarter of a millimetre thick, whose surfaces are ground and polished to as smooth a finish as

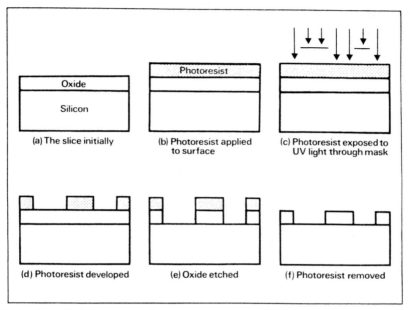

44 A schematical illustration of the process of producing a silicon chip. (Courtesy of the Science Museum, London)

possible, in order to remove as many defects as possible. Depending on their size, about 500 chips are laid side by side on a wafer. Each chip contains about a million components or junctions. These junctions consist of negative and positive charge carriers, in which electron 'holes' are positive (p types) and excess electrons are negative (n types) (*see* Figure 43). Used together, these junctions create all manner of electronic components such as transistors, diodes, and capacitors. The p and n junctions are produced by a highly selective doping procedure in which, for example, boron atoms in the silicon create the electron holes and phosphorus produces the excess electrons. As illustrated in Figure 44, this is achieved by coating the wafers with silicon dioxide, and then with a photosensitive material. The circuitry pattern required is then exposed photographically onto the photosensitive coating, and permanently imprinted by hardening the layer. The unhardened layer is finally removed by an acid leaving the pure silicon base unaffected. Doping is then carried out by exposure to an atmosphere of phosphorus or boron at a temperature sufficiently high to cause the atoms to diffuse from the atmosphere through the 'windows' in the silicon

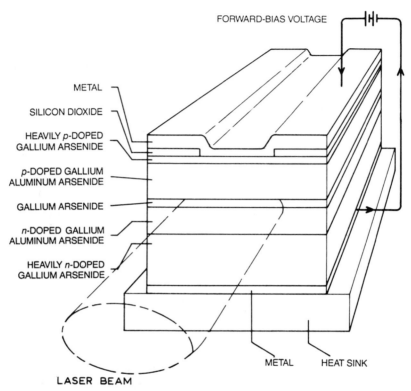

FORWARD-BIAS VOLTAGE

METAL

SILICON DIOXIDE

HEAVILY *p*-DOPED
GALLIUM ARSENIDE

p-DOPED GALLIUM
ALUMINIUM ARSENIDE

GALLIUM ARSENIDE

n-DOPED GALLIUM
ALUMINIUM ARSENIDE

HEAVILY *n*-DOPED
GALLIUM ARSENIDE

METAL   HEAT SINK

LASER BEAM

45   A gallium arsenide heterojunction laser device. The various material layers, deposited by physical vapour deposition, effectively confine emitted light to a narrow slit. (J. M. Rowell, *Scientific American*, Oct. 1986)

dioxide layer produced by the masking process. The production of silicon chips is a very demanding one, and nowadays up to about 50% of all chips produced are rejected because of the presence of small atom-sized defects in the silicon lattice, as well as other problems.

After silicon

As we have seen, a problem with silicon chips is that the energy needed to bring about the flow of electrons causes the chip to heat up, and that as a result in a large computer measures must be taken to avoid overheating. A better material in this respect is gallium arsenide (GaAs), which has the advantage that less energy is required to induce flow, so there is less of a heating effect than in doped silicon. As noted in conjunction with Figure 43, the band gap of a semiconductor defines the energy needed to raise electrons into conduction states or, conversely, the energy given out when electrons move out of conduction states. In compound semiconductors like GaAs this energy is usually in the form of light waves. Devices which can exploit this type of two-way conversion of energy between light waves and electric currents include lasers and solar cells. An example of the way gallium arsenide laser devices are constructed is shown in Figure 45.

Gallium arsenide has a much higher electron mobility than silicon, and as such it is a better material than silicon for use in low-energy high-frequency applications such as fibre-optical lines, remote control devices for TV, and car telephones. A disadvantage, for the materials scientist, is that gallium arsenide is more difficult to produce than silicon in a form that is pure and free of defects. An alternative is to evaporate thin, defect free, films of GaAs onto a silicon substrate material. Chips of this composite material have been produced in which islands of GaAs are employed as ultrasensitive optical switches within the silicon semiconductor circuit, thus combining the best qualities of both materials. It is predicted that the world market for gallium arsenide semiconductor devices will increase from the present-day 200 million dollars to around 4 billion dollars by the turn of the century.

Gallium arsenide is not of course the only new potential semiconductor material being examined. The glass-based selenium–germanium alloys are an interesting example of how amorphous materials can be used as semiconductor materials. When a pulse of electrical energy is applied to this material, the glass heats up to a temperature just above the glass transition temperature, causing the glass to crystallise locally in the form of conducting fibres that span the amorphous matrix film. These fibres are so fine that they contain no defects, so conduction along them is extremely efficient. The applied electrical pulse thus has the effect of rapidly increasing the conductivity of the material. When a second pulse is applied the crystalline fibres melt and then cool so rapidly that they revert to the glass or amorphous structure, with the result that the device returns to its former low-conductivity state.

These amorphous semiconductor materials are currently being considered for use in computer memories as erasable 'read-only' devices. It is also possible that the energy pulses needed to bring about the crystallising transition could be supplied by a laser. It may therefore be feasible in the near future to use these simple types of semiconductor in transmitter/receiver devices in fibre-optical telecommunication systems.

Conducting polymers and plastic chips

Today, semiconducting polymers are used in a variety of applications including electronic and electrical components such as switch contacts, resistors, and electrodes, and as guards against lightning and electrical discharge. Apart from electrical conductivity, the associated property of good thermal conductivity is also useful for obtaining rapid through-heating of parts produced in extrusion and forming operations.

Two ways have so far been used to make polymers electrically and thermally conducting. One is simply to embed a fine dispersion of conductive fibres or particles into the plastic; the other is to paint a conductive coating on the surface of the part. However, a more fundamental approach is to modify the conduction bands of the polymer molecules by alloying or doping.

The use of doping to increase conductivity is a recent innovation of scientists and provides the key to the development of the plastic chip. A dopant such as arsenic pentafluoride is added to a suitable polymer such as polyacetylene. As in silicon semiconductors, the effect of doping is to modify the conduction bands of the polymer molecules so as to bring about electron acceptor/donor manipulations. For example, by exposing polymers such as polyacetylene to an iodine vapour the polymer is made to give up electrons to the vapour phase, increasing the mobility of electrons in the polymer; in other words it becomes a conductor. Another approach is to dope graphite fibres with iodine or bromine. This causes ions to form which help transfer charged particles along the fibres. However, there are problems with producing conducting polymers in these ways. Doped polymers are not very stable in air, and the dopant quickly leaks away. A conducting polymer that is both cheap to produce and stable in air is needed if plastic chips are to be a realistic proposition. Such a polymer does not exist today, though given the present level of research activity in this area the cheap plastic chip will probably appear by the turn of the century.

In summary, silicon semiconductor chips are likely to be displaced, or maybe even superseded, by chips made from gallium arsenide or even plastic. On this basis, tomorrow's computers and electronic de-

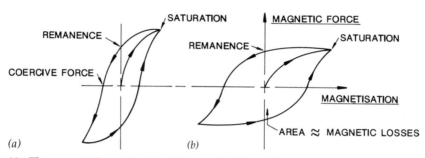

46 The magnetic loss or 'hysteresis' curves for (*a*) a soft magnet and (*b*) a hard magnet. The significant difference is that soft magnets easily lose their magnetisation, whereas hard magnets remain magnetic. These characteristics determine their uses

vices will operate more rapidly, need less energy, and require less cooling than the machines we use today. Even the elements connecting individual chips will be more efficient in future, as they will probably be based on optical fibres, or even superconductors, which not only transmit faster than normal metal conductors, but also will not impose the high capacitive loads that slow down and heat up today's computers.

Magnets

Magnets are to be found all around the home: they energise the electrical motors in vacuum cleaners and refrigerators, they help reproduce sound and visual images in stereo and video equipment and store information in computers, and they keep cupboard doors closed. Indeed, man has used magnets – or at least known of them – for at least 3000 years, thanks to the abundance of the naturally magnetic mineral magnetite, commonly known as lodestone. By the sixteenth century it had been demonstrated that if freshly smelted iron is drawn in the shape of a bar, it becomes magnetic.

From the technological point of view an ideal magnetic material should have one of two properties: it should be either 'soft', meaning easy to magnetise or demagnetise, or 'hard', meaning that it remains magnetic or is a permanent magnet. Whether a magnet is hard or soft is decided by whether it remains magnetic after an externally applied electrical field has been removed. This degree of magnetisation, in terms of the magnet's remanence and coercivity, is illustrated in Figure 46, and is often referred to as the magnet's 'hysterisis' after the Greek

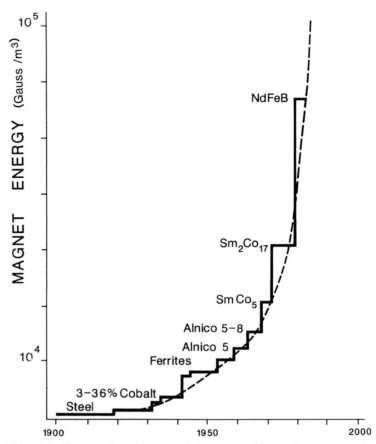

47 This graph illustrates how the strength (as measured by maximum magnetic energy) of permanent magnets has increased with the development of new materials. The powerful new magnets will be used in a wide variety of applications, such as lightweight headphones and car telephones

word meaning 'to lag'. Generators, electric motors, and transformers operate at their maximum efficiency if magnetisation does not remain after the external field has dropped to zero, and this requires a soft magnet of low remanence with the type of hysterisis loop shown in Figure 46(*a*). On the other hand, permanent magnets, used in refrigerators, doorcatches, and earphones, need to have a higher remanence with a hysterisis loop of the kind shown in Figure 46(*b*).

The best traditional permanent magnetic materials are based on 'ferrites'. Over the years, however, alloy development has resulted in substantially stronger (harder) magnetic materials than the ferrites, as illustrated in Figure 47. In more recent years, hard magnets have been made from sophisticated alloys that can be produced in a particulate form, such that each particle effectively represents a single magnetic 'domain'. These individual magnetic particles can then be aligned under an applied electrical field and subsequently sintered or mixed with a suitable binding material to make very powerful magnets indeed. In another approach, materials such as an iron–boron alloy, with additions of silicon or neodymium, are formed into amorphous ribbons by very rapid solidification. These materials show extremely low hysteresis losses when magnetised, which means that the magnet is very soft and remagnetisation can occur any number of times with low losses. Such materials are now being used as laminated core material in transformers and electric motors, and for audio tape recording heads.

Other applications of magnetic materials call for properties that lie halfway between hard and soft magnets. The magnetic memory element in a digital computer, for example, must be hard enough to retain its forward or reverse magnetisation ('zero' and 'one' states in the computer's binary system), yet soft enough to switch states cleanly and rapidly when a small external load is applied during read-in or read-out procedures. For this purpose, tailor-made materials are being produced based on 'bubble memory' characteristics which change their properties in a super-sensitive way, not unlike some of the functions of the human brain. However the future may again be in the hands of polymer chemists. Soviet researchers, for example, have recently produced 'ferromagnets' based entirely on polymers (polydiacetylene), with potential for a wide range of new magnetic materials.

Superconductors

If certain materials are cooled to temperatures of some 250°C below the freezing point of ice, something almost miraculous occurs: they become superconducting – able to conduct electricity with virtually no loss of energy in the process. Normal electrical transmission lines, for example, have energy losses of 15–20%, so that if only superconducting cables could be used at reasonably high temperatures ('high' here means above the temperature of liquid air, −196°C), this would be a major breakthrough with great commercial significance in power transmission and many other applications. The liquid air temperature,

Selected wonders of the world of advanced materials 5
'Warm' superconducting oxides

These materials are remarkable in that they possess zero resistance to the flow of electrical current at a relatively 'warm' temperature, above that of liquid air. The most reliable and technologically interesting of these oxides is based on a copper oxide containing additions of yttrium and barium. Oxides or carbonates of the constituents are ball milled into extremely fine particles, cold pressed, and then fired (or sintered) at around 925 °C. By cooling the partially sintered compact slowly, the lattice structure takes up oxygen and an unusual distorted perovskite-type single phase oxide is formed with the composition $YBa_2Cu_3O_7$. This material is metallic in appearance and properties and has in fact an ambient temperature conductivity almost as low as that of copper. Most remarkable of all, however, is that if it is cooled to a temperature of about 90 K (*ca* 23 °C above the temperature of liquid air) it becomes a superconductor. Since it is easy (and inexpensive) to produce liquid air, these superconductors have enormous potential use, for example, as zero-loss power transmission lines, super-strength magnetic materials, and materials for levitating trains. The figure demonstrates the fine microstructure of this superconducting oxide. The striations across the elongated crystals are thin 'twins' or faults caused by the transformation from the high temperature tetragonal (non-superconducting) phase to the low temperature orthorhombic (superconducting) phase.

Light optical micrograph by J. P. Zhou, University of New South Wales
(Magnification × 325)

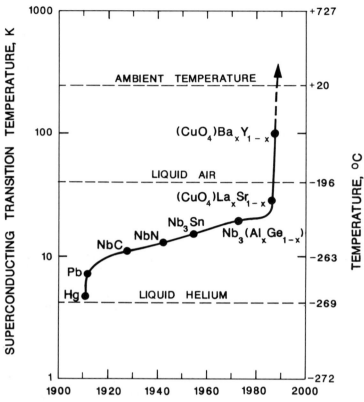

48 The development of superconducting alloys and ceramics since the discovery of the phenomenon in 1911. The transition temperature is the temperature below which the conductor has zero resistance to the flow of current

−196°C, may sound cold – and hence expensive to create – to some people (it is about three times lower than the mid-winter temperature at the South Pole), but in fact liquid air is relatively cheap to manufacture, costing about the same as beer. Recently, such a breakthrough in 'high-temperature' superconductivity appears to have been made, with the discovery that certain ceramic materials based on copper oxide are superconducting at temperatures well above the freezing point of air. Quite suddenly, the realm of superconducting transmission lines, miniature high-speed computers, and supermagnets is on our doorstep.

The discovery of superconductivity resulted from experiments by Heike Onnes, a Dutch scientist at the University of Leiden in 1911, who

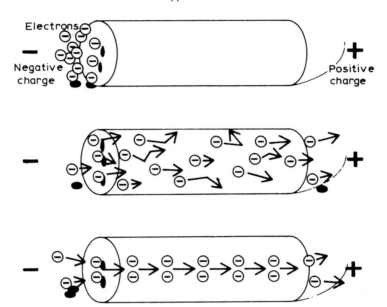

49 The flow of electrons in an insulator, a normal conductor, and a superconductor. (From *Time*, 11 May 1987)

found that the electrical resistance of mercury decreased as expected on cooling, but on reaching a temperature of 4·15 K (about −269°C) its resistivity suddenly vanished abruptly and completely. As illustrated in Figure 48, since that time – and despite much research – little progress was achieved in finding a material that loses its resistivity above liquid air temperature. Until 1986 the record transition temperature, set by a metallic niobium–germanium compound, was only −250°C. The breakthrough came about in that year when researchers at the IBM laboratories in Zurich achieved a superconducting temperature of −237°C in a copper oxide compound containing minor additions of lanthanum and strontium. That superconductivity should be achieved in a ceramic, the most traditional of insulating materials, was amazing in itself and immediately provoked a new wave of interest in the subject.

We still do not understand why these or any other materials are superconducting, so obviously it is difficult to say why ceramic materials should be better superconductors than others. The phenomenon occurs when the normal process of scattering between electrons and the compound's lattice are suddenly removed. The materials science of

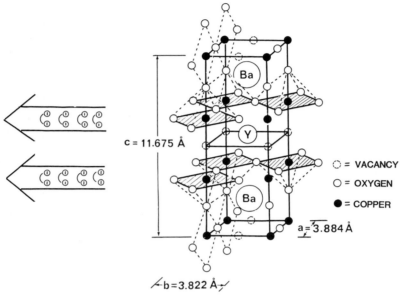

c = 11.675 Å

○ = VACANCY

○ = OXYGEN

● = COPPER

a = 3.884 Å

b = 3.822 Å

50   A unit cell of the superconducting ceramic $YBa_2Cu_3O_7$. The structure is
orthorhombic in form and the 'vacancies' refer to 'missing' oxygen atoms.
Superconducting planes are shown shaded and the direction of superconduction
is indicated by arrows

this is complex and interesting. Superconduction is thought to be
associated with superfine-layered structures of just one or two atoms
thick, which somehow distort the lattice and amend its valency bands
sufficiently to bring about zero-resistance free 'tunnels' through which
electrons or electron pairs move unimpeded. The difference between
superconducting, normal conducting, and insulating behaviour is
illustrated in Figure 49.

   The electron collisions (Figure 49) which occur in normal conduction
are due to interactions between the moving free (valance) electrons
and the phonons. Phonons are quantised wave phenomena brought
about by the lattice vibrations of the atomic structure. It is conjectured
that in superconduction there are no collisions between the electrons
and phonons; instead, the electron pairs that form up move in unison
with the lattice vibrations so as to assist the electrons along their paths.
Normal electron conduction could be compared with paddling out to
sea on a surf board, while superconduction is like riding the board back
on the incoming waves.

In the case of the new copper oxide based superconductors, it is found that addition of rare earth elements such as lanthanum, strontium, scandium, and yttrium to a compound of copper and barium oxide completely changes the normally insulating properties of this oxide to produce a superconductor. Indeed the material is a fairly good conductor even at room temperature, with properties and appearance not unlike that of a typical metal. The most successful and stable of these superconducting oxides has the atomic composition $YBa_2Cu_3O_7$. This compound exhibits zero resistivity and superconducting properties below a temperature of $-175°C$, some 22°C *above* the temperature of liquid nitrogen (*see* Figure 48). Materials scientists are currently experimenting with compositional modifications to further increase the zero resistivity temperature. In terms of overall stability and repeatability during temperature cycling, the $YBa_2Cu_3O_7$ material is currently the best candidate as far as technological application is concerned. However, new and better materials may be on their way.

An example of the atomic configuration of this compound is shown in Figure 50. It is seen that the structure is layered in the sense that it contains conducting planes dense in copper and oxygen atoms (shown shaded), separated by non-conducting planes corresponding to the positions of the yttrium and barium atoms. Superconduction occurs along these copper and oxygen electron-rich planes as indicated by the arrows.

Making power cables out of these ceramic compounds will certainly be a task to tax the ingenuity of materials scientists. For most applications the superconducting wire ought ideally to be a good 'normal' conductor at ambient temperature, otherwise there is the risk of the cable being destroyed by sudden heating as it warms up above the superconducting transition temperature. To achieve this it may be necessary to integrate the superconducting compound with, for example, normal copper cable, unless compounds retaining their superconductivity up to ambient temperatures can be produced.

Potential applications of superconductors are many. Power transmission lines have already been mentioned. It would be easier to lay these underground, under cities for example, than it would normal cables, which tend to have high energy losses and suffer from heating problems. Schemes for cooling underground superconducting cables have already been advanced, including passing cold gases over the superconducting lines within sealed covers.

In computers, superconducting switches can flip at least as fast as

51 'Levitation' – the magnetic field of a magnet is repelled by a superconductor causing the magnet to hover above the superconductor. This is known as the Meissner effect

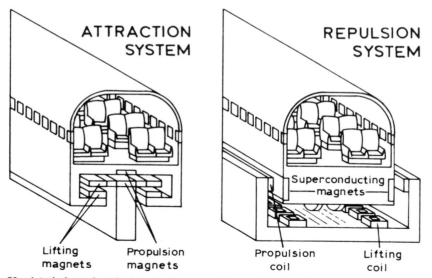

52 A train is made to levitate by the powerful repulsive force between superconducting magnets in the train and magnetic 'railway lines'. (From *Time*, 11 May 1987)

gallium arsenide chips, and when used as a circuit component a super-conductor can communicate with other components faster than normal conductors. In addition, superconductors dissipate practically no heat and chips can be packed tightly together, making further miniaturisation possible. In medicine, metal superconductors operating at liquid helium temperatures ($-269°C$) are already used in nuclear magnetic resonance scanners. These scanners produce images showing the concentration of certain atoms in the body, and hence provide information on the function and condition of the body. For instance, they can help to locate the presence of small tumours. Very high magnetic fields are normally needed for this technique, requiring superconducting magnets cooled to liquid helium temperature. Super-conducting oxide magnets operating at higher temperatures need only be cooled by liquid air or carbon dioxide. With the help of the new superconducting ceramics, it is quite conceivable that a nuclear magnetic resonance scanner of much simpler construction could well be in every specialist's consulting room by the turn of the century.

One of the unique characteristics of superconducting materials is that they repel ordinary magnetic fields. This gives rise to a kind of levitation (*see* Figure 51) in which, for example, superconducting materials could be used to keep trains clear of magnetised rails, as illustrated in Figure 52.

There are many other ways in which high-temperature superconductors may affect our lives. As with the development and application of other advanced materials, the ways in which these remarkable materials are put to the service of mankind is limited only by the imagination and conscience of the citizens of tomorrow's world.

Further reading

Popular accounts of the applications of advanced materials are to be found in popular scientific and technical magazines such as *New Scientist* and *Scientific American*, as well as more specialist journals such as *Advanced Materials and Processes* (published by ASM International). Particularly recommended is the October 1986 issue of *Scientific American*, which is devoted entirely to new materials.

GLOSSARY

alloy, alloying   An alloy is formed by the mutual solution within a single phase or crystal structure of two or more elements. Even in alloys where several phases coexist, all the phases present are likely to be *alloyed*, i.e. contain two or more constituent elements.

amorphous   Material with a random or non-symmetrical atomic arrangement. In metals and ceramics it means a *non-crystalline* state; in polymers it means a random arrangement of the atomic chains of the polymer.

anisotropic   This refers to asymmetrical properties or shape of a crystal lattice or cellular structure. Thus the individual cells of wood are asymmetrical in shape, and have different stiffnesses when loaded in different directions.

austenite, austenitic   The high-temperature phase, or (face centred cubic) crystal structure of iron and steel. When steel is hardened, it must first be annealed in the austenitic condition where its carbon content is easily dissolved.

binary alloy   An alloy with two constituents. For example, steel is a binary alloy of iron and carbon.

biodegradable   This refers to synthetic polymeric materials or compounds which are capable of disintegrating to some natural form or substance.

biomaterial   Synthetic material used for implants in the human (and animal) body.

blend   A polymer material containing two or more constituents within a given phase. Blends are also referred to as *polymer alloys*.

carbon steel   All steels contain carbon (the definition of steel is that it is an alloy of iron and carbon), but carbon steel is usually of a ferritic form at ambient temperature.

cell   In materials science, an open or closed symmetrical unit of a cellular material such as the single cell of a bee's honeycomb, or of wood.

cermet   A composite of ceramic and metal, such as silicon carbide fibres in aluminium.

composite   A material consisting of at least two separate phases or

105

components such that the whole has properties superior to the individual components. Examples are reinforced concrete and glass-fibre-reinforced polymers.

co-polymer   A composite of two or more polymer types.

creep   A form of *plastic deformation*, usually occurring at quite a high temperature and over a long period of time.

crystalline   On solidifying, the atoms of many materials form up in an almost perfect symmetrical lattice structure. Almost all metals are crystalline; other classical examples of crystals are diamond and ice.

deep drawing   A manufacturing technique in which metal is shaped by drawing in a special die to its final form.

dislocation   A linear discontinuity in the perfection of a crystal. A sufficiently large external force acting on the crystal may cause the dislocation, or groups of dislocations, to move, and this is the basis of *plastic deformation*.

dopant, doping   A very small amount of an alloying addition which, when dissolved by a material, may modify its properties. An example is the doping of silicon by phosphorus or boron to make it semiconducting.

equilibrium   The stable state of a phase or structure. These stable states are illustrated for most binary and many ternary alloys in the form of equilibrium diagrams.

extrude   A metalworking process in which metal is forced through a die or orifice.

fatigue   A form of failure or fracture in a material produced by cyclic loading which can occur at loads well below the normal (static) strength of the material.

ferrite   The low-temperature (body centred cubic) phase of iron. In most steels it is the stable phase at ambient temperature.

fine-grained microstructure   Most crystalline materials are *polycrystalline*, comprising of many small crystals separated by crystal or grain boundaries. A fine-grained microstructure is a polycrystalline material with a fine grain size.

glass transition temperature   A term used in polymer science for the temperature at which a change in stiffness or modulus of the material occurs. These polymers should not normally be used above the transition temperature.

grain   An individual crystal within *polycrystalline* (multi-grain) material, in which adjacent grains have substantially different orientations.

grain boundary   Each grain in a polycrystalline material is separated by a grain boundary about two atoms wide.

hardness   An intrinsic property of a material, usually measured by an indentor of a specific shape and loading.

heat affected zone   When metal is fusion welded, the areas near the hot part of the weld metal will be heated. This heating cycle can cause substantial changes in microstructure (e.g. grain size) and properties.

hot isostatic pressing   A process for densifying ceramic or metallic powders under conditions of high temperature and isostatic pressure.

impact toughness   *Toughness* refers to a material's resistance to crack growth. Impact toughness is its crack growth resistance under conditions of impact or rapid loading.

injection moulded   Often refers to the shaping or forming of a *thermoplastic* material by injection through a suitable die at elevated temperature.

interstitial   A non-metallic atom in an alloy or compound small enough to occupy sites in the lattice between the metal atoms. In a dilute solution of interstitials in a metal, such as carbon in iron, interstitial atoms can migrate (diffuse) without the need for *vacancies*.

ion implantation   A sophisticated process in which ions are impacted into the surface of samples causing modifications in properties of the surface layers. Since the process occurs at low temperatures the implanted ions cause extreme distortion in the lattice structure of the substrate material, greatly affecting its properties.

lattice   The geometrically perfect network structure on which crystalline materials are based.

martensite, martensitic   The phase resulting from a diffusionless (martensitic) phase transformation. In steel martensite is a solid solution of carbon in iron.

matrix   The dominating lattice structure of a given phase in a material.

modulus   A constant of the elastic strength of a material. For example, Young's modulus (E) refers to the constant of elasticity, or stiffness of a material.

order   This usually refers to the structure of an alloy of *stoichiometric* composition. For example, compounds such as aluminium oxide (Al_2O_3) or titanium nitride (TiN) are perfectly *ordered*, each component species forming its own lattice structure, and with the combination forming a *superlattice* structure. It should be noted, however, that in the present text *order* is sometimes used to describe the symmetrical crystalline state of a metal, as opposed to an unordered amorphous state.

phase   A microstructural constituent of one given lattice structure or type. A material may contain from one to several coexisting phases.

physical vapour deposition   A process for depositing extremely thin layers of material onto a substrate through a vapour or gaseous phase passing over it.

plasma spraying   A process for spraying materials onto a substrate or surface in which the materials, in powder form, are heated in a hot plasma created by an electric arc. The technique is particularly useful for laying down surface coatings of ceramic materials.

plastic deformation   An irreversible change of shape of a material, brought about in crystalline materials by the creation and movement of large numbers of *dislocations*.

polycrystalline   This refers to a crystalline material containing several (usually very many) grains.

precipitate   A minor phase or particle that has precipitated by the movement and coalescence of individual atoms from solid solution during heat treatment.

precipitation-hardened   This refers to a material in which the precipitation reaction has been so finely dispersed that the material becomes hardened. The small particles provide a barrier to dislocation movement during *plastic deformation*.

segregation   In this process alloying elements diffuse out of *solid solution* and collect at defects in the material such as grain boundaries.

sintering   The fusion of metallic or ceramic powders into a dense whole when held at elevated temperature. The process occurs in the solid state by interparticle diffusion.

solid solution   Very few practical materials are pure in form or consist of only one element. If one element or more are dissolved completely in a base material in the solid state, then the whole is referred to as a solid solution.

stress corrosion   A combination of mechanical stress and corrosion which causes an accelerated form of corrosion in the metal.

substitutional   This refers to atoms of an alloying addition to a metal which occupy, or substitute, positions or atoms of the matrix material.

superalloys   Multiphase alloys, usually cobalt- or nickel-based, which are very heat resistant.

superconductor   A material which has practically zero resistivity. The phenomenon, which is thought to be associated with a concerted movement of coupled electrons, occurs in only a few materials and only below some critical temperature.

thermoplastic   A polymer that can be plastically shaped or worked at elevated temperature, hence thermo-*plastic*. Polyester and PVC are examples.

thermoset   A polymer that cannot be plastically shaped or worked at elevated temperature. It is thus thermo-*set* (in shape). Epoxy is an example.

toughness   A property of a material describing its degree of resistance to crack growth or fracture.

transformation   The transition from one crystalline phase to another. A phase transformation can be brought about either by a diffusion-controlled reaction or by a diffusionless (martensitic) shear process. The transformation is brought about essentially by chemical changes in energy of atomic bonding that occur on changing the temperature.

vacancy   A 'point defect' in a crystal, an atom missing from the lattice structure. It has the important function of aiding the process of diffusion.

weldability   A material has good weldability when it can be joined by fusion welding without problems of cracking or excessive porosity occurring.